LIME ROCK PARK

35 YEARS OF RACING

PROGRAMS
18

LIME ROCK PARK

35 YEARS OF RACING

BY RICH TAYLOR

FOREWORD BY SAM POSEY

Sharon Mountain Press

Lime Rock Park: 35 Years of Racing

Every effort has been made to make the information presented in this publication as accurate as possible and the opinions expressed as well-informed as possible. Lime Rock Park, the author and publisher regret any errors of fact, omission or interpretation. They are assuredly not intentional.

For information: Sharon Mountain Press, Inc., 1 Jewett Hill Road, Sharon, CT 06069.

Book Design by Jean Constantine.

Taylor, Rich
Lime Rock Park: 35 Years of Racing
Rich Taylor
ISBN 0-9633994-0-3
1. Lime Rock Park—History. I. Title.
Library of Congress Catalog Card Number: 92-064436
First Edition

Contents

Acknowledgements

(Above) BMW 320i in IMSA GT trim stops in the Lime Rock pit lane, 1977.

A book is like a racing car; one person's name is on the side, but it takes a whole team to make it go. Phil Rosette, Vice-President of Marketing at Lime Rock Park, originally came up with the idea to publish a history of the track and did everything he could to help it run smoothly.

Skip Barber, President of Lime Rock Park, encouraged me, as did Mike Rand, Vice-President of Operations and General Manager. Other people at Lime Rock Park who helped me immeasurably include Donna Maxwell, Rita Kurek, Butch Sherwood, Dick Bayer, David Hall and Historians Jan Poor and the late Connie Lovejoy.

Dozens of racing personalities graciously talked to me, in general about American road racing over the past four decades and in particular about Lime Rock Park. Some spoke for just a few moments, some reminisced for literally days at a time. I owe them all a great deal, including anyone I may have momentarily forgotten.

Among them are Bob Akin, George Alderman, Gaston Andrey, Skip Barber, Kenny Brenn, Don Breslauer, Tim Coconis, Joe Corbett, Newt Davis, Alex Dearborn, Jack Dolan, Chris Economaki, Joel Finn, John Fitch, Charlie Gibson, Jim Haynes, Ed Jacobs, Rick Kopec, Malcolm Labatt-Simon, Tony LaPalme, Bob Leitzinger, Sandra Leitzinger, Bruce MacInnes, Gordon MacKenzie, Charles Mallory, Bill Mayberry, Denise McCluggage, David Moore, Mary Moore, Nick Moore, Eric Morrow, Karen Petersen, Sam Posey, Brian Redman, Greg Rickes, Craig Robertson, Alex Saidel, Judi Chatfield, Jim Shane, Bob Sharp, Scott Sharp, W. B. "Fred" Stevenson, Judy Stropus, Henryk Szamota, Harry Theodoracopulos, Dick Thompson, Rodger Ward and the man who started it all, Jim Vaill.

Sam Posey, Skip Barber, Greg Rickes, Phil Rosette, Mike Rand and Donna Maxwell proofread the manuscript for correctness. I owe them an extra thank you, particularly Greg Rickes. I'd also like to thank Bruce MacInnes and Carl Lopez for help in writing "One Lap of Lime Rock" and Sam Posey for writing the Foreword with his usual wit and charm.

I'd also like to thank the hundreds of fellow competitors, workers and officials from the SCCA, IMSA and VSCCA who have helped me enjoy so many memorable days of racing at Lime Rock over the past two decades...and helped me survive a few afternoons I'd rather forget.

Most of all, I want to thank my wife and partner, Jean Constantine, who designed this book, did the layouts and mechanicals, collected the photos, did typesetting, copy-editing and historical research. She also endured my creative tantrums with understanding and good humor and met impossible deadlines with grace under pressure. I'll make it all up to her some day, I promise.

Rich Taylor
Sharon, CT
July 17, 1992

Foreword

My first ever view of Lime Rock was from the grassy hill overlooking the first turn from the outside of the track. I was 13 and already in love with racing, but my contact with the sport was strictly through books. Now here I was actually looking at D-Jaguars and Maseratis (it was 1957). Real racing cars, almost in my back yard. Lime Rock was like a miracle.

It was also very green, as it still is today. Dante used green as the symbol for hope in *The Divine Comedy*, and in my daydreams Lime Rock soon became fused with my emerging obsession for racing. Notice that I said my first view of the track was from the outside—as a fat kid, no good in school, I saw life from the outside. But from the hill, the world I so hoped to become part of seemed for the first time to be within my reach.

For a modest sum, my mother became a stockholder at the track, and with a lot of generous-spirited people looking the other way I was soon out on the track itself, first with a go-kart, later with a Formula Vee purchased from John Fitch. I ran my first race at Lime Rock, and won my first race there, too. Best of all, I became part of an unofficial Lime Rock family, which consisted not only of the people who ran the track but also certain hard-core fans. The late "Spanky" Smith, for example, was a fixture at the track along with his unique motor home (decorated inside and out with racing memorabilia), and one year, when I arrived exhausted from the Monaco Grand Prix, he had the bed made up and my favorite meal waiting for me.

I drive to the track along the back road from Sharon, and as I approach I always feel my love for the place mixing with apprehension. In the fall of 1967, testing a Caldwell Can-Am car, I crashed on the uphill, going end over end through an oak, which to this day has pop rivets from my chassis embedded in it about twenty feet up. I was trapped in the car and unconscious for a while, and when I got out I wandered around the woods in shock. At home that night I thought, "This place can kill me," and I've never forgotten that.

(Above) A young and happy Sam Posey, having just cut the first lap at Lime Rock under 1 minute. He did a 58.0 in his McLaren on July 4, 1967.

Dan Gurney brought Juan Fangio II to our house for dinner the night Juan won the Camel GT race, and Juan spoke with passion about the track, calling it mysterious, a riddle he couldn't solve, a track where a perfect lap will never be driven.

I agree. The faster you go, the clearer it is that something inherent in the geometry of the track guarantees that speed gained in one place hurts you somewhere else; that the alternation of containment and release of the car's energies have an unsolvable overlap.

In my fantasies, a floating, almost lyrical lap seems possible, but the reality is much more physical. You battle the track. You explode under the bridge headed for the downhill, tipped toward the earth like a doomed fighter pilot ("Point your toes, boys, we're going in!"). One lap propels you into the next, but, at the limit, the rhythm you seek is not to be found.

Today, thanks to Skip Barber having allowed my brother David and myself to be the architects of the track's new buildings, it is possible for me to stand on the porch of a building I helped design, in the exact spot I first saw the track 35 years ago. I find the new Jags as exciting as the D-types were then, and if the Maseratis are gone, well, we have the Nissans and Toyotas to watch as they slice through the greenery. More fans than ever now crowd the hills. For all of us, Lime Rock continues to be a very special place.

Sam Posey

Photo Credits

Our sincere thanks to all of you who generously shared your photos and memorabilia; we couldn't have done it without your gracious help.

Please note: all photos are listed by pages from top to bottom and left to right.

DUST JACKET

front cover: background LRP photo file, Gordon MacKenzie, Rich Taylor, Randy McKee, Randy McKee. author: Rush Workman.

END PAPERS

front: LRP photo file, back: Ed Jacobs.

FRONT MATTER

p i: Paul Gulash, p ii: LRP photo file; Randy McKee, p vi: Andrew Hartwell, p vii: Sam Posey, xi: Richard Pocock.

PROLOGUE

p 1: Rich Taylor, p 3: John Fitch, p 4: Rich Taylor, p 5: John Fitch.

CHAPTER 1

p 7: Gordon MacKenzie, p 8: map drawn by Jean Constantine, p 9: Peter Stettner, clip art, Peter Stettner, p 10: Jim Vaill, p 11: Jim Vaill, p 12: Jim Vaill, p 13: Jim Vaill, p 14-15: Penny Havard; p 16: John Fitch, p 18: Gordon MacKenzie, p 19: John Fitch, Gordon MacKenzie, p 20: both John Fitch, p 21: John Fitch; Stephen Dunn, John Fitch; Bill Hughes, John Fitch, LRP photo file, p 22: Judi Chatfield, p 23: Jim Vaill, LRP photo file, p 24: Judi Chatfield, Gordon MacKenzie, p 25: Judi Chatfield, LRP photo file.

CHAPTER 2

p 27: Sandra Leitzenger, p 28: Judi Chatfield, p 29: Judi Chatfield, AAVR, Inc. Robert F. Williams, p 31: LRP photo file; Fred Rosvold, Judi Chatfield, p 32: LRP photo file, Robert Couch, p 33: Rene Dreyfus, AAVR, Inc. Robert F. Williams, AAVR, Inc. Robert F. Williams, Bruce MacWilliams, p 35: Robert Gorman, Judi Chatfield, Rene Dreyfus, AAVR, Inc. Robert F. Williams, Robert Couch, p 36: Judi Chatfield, Alex Saidel, p 37: Alex Saidel, Judi Chatfield, Alex Saidel, Judi Chatfield, p 39: John Fitch, p 41: background Tony Thurston, Judi Chatfield, Judi Chatfield, Judi Chatfield, p 43: background Judi Chatfield, AAVR, Inc. Robert F. Williams, AAVR, Inc. Robert F. Williams, AAVR, Inc. Robert F. Williams, p 44: AAVR, Inc. Robert F. Williams, p 45: Judi Chatfield, p 46: John Fitch, p 47: both Gordon MacKenzie, p 48-49: Gordon MacKenzie, p 51: AAVR, Inc. Robert F. Williams, Judi Chatfield, p 52: Dave Pettigrew, p 53: Newt Davis, Judi Chatfield, Newt Davis, p 55: AAVR, Inc. Robert F. Williams, Greg Rickes, p 56: Judi Chatfield, p 57: AAVR, Inc. Robert F. Williams, Judi Chatfield, p 58: Harvey Siegelman, p 59: LRP photo file; Randy McKee, LRP photo file; Phil Parsons, p 61: Judi Chatfield, Judi Chatfield, LRP photo file, p 62: AAVR, Inc. Robert F. Williams, p 63: all AAVR, Inc. Robert F. Williams, p 65: both LRP photo file.

CHAPTER 3

p 67: Bill Mayberry; Bill Warner, p 69: Joe Corbett, LRP photo file, p 71: LRP photo file; illustration by Sam Posey, Joe Corbett, p 72: LRP photo file, p 73: LRP photo file; John Peckham, LRP photo file; Jerry Milks, LRP photo file; Phil Parsons, p 75: LRP photo file, Joe Corbett, LRP photo file, p 77: LRP photo file; drawing by John Peckham, p 78: Bill Mayberry, p 79: LRP photo file; Randy McKee, LRP photo file, p 81: Bill Mayberry; WC Baker, p 82: all Greg Rickes, p 83: LRP photo file; Gary Van Voorhis, LRP photo file, p 84: Greg Rickes, LRP photo file; Gary Van Voorhis, Bill Mayberry, p 85: LRP photo file; Thomas Maroney, p 87: Bill Mayberry, p 88: Greg Rickes, p 89: Ed Jacobs, Bill Mayberry, Sam Posey, p 90: Rick Kopec, Greg Rickes, p 91: JVS Enterprises, Russ von Sauers; Spanky Smith, p 92: John Vorisek, Denis Tanney, p 93: Greg Rickes, LRP photo file; Ellery Kington, p 94-95: Rich Taylor, p 97: LRP photo file, p 99: LRP photo file; Jim Fisk.

CHAPTER 4

p 101: Ed Jacobs, p 103: Lew Kerr, p 104: LRP photo file, p 105: Joe Corbett, Bill Mayberry; Bill Oursler, p 107: Bill Mayberry, Russ von Sauers; Spanky Smith, p 108: Dan Yasek, p 109: Greg Rickes, p 110: Russ von Sauers; Spanky Smith, Ed Jacobs, p 111: Ed Jacobs, Greg Rickes, p 112-113: Ed Jacobs, p 114: LRP photo file, p 115: background LRP photo file; Fred Bull, LRP photo file; Randy McKee, LRP photo file, LRP photo file; Randy McKee, LRP photo file; Mike Miller, p 117: LRP photo file; Randy McKee, Dennis Yuro, p 119: Mike Rand, p 120: Richard Niezabitowski, p 121: LRP photo file; Bridge Hill Studio, LRP photo file, LRP photo file; Swede Plaut, p 122: Bill Mayberry, LRP photo file, Richard Niezabitowski, p 123: Ed Jacobs, p 125: LRP photo file; Geoffrey Hewitt, p 127: LRP photo file, Ed Jacobs, Greg Rickes.

CHAPTER 5

p 129: LRP photo file; Randy McKee, p 130: Ed Jacobs, Tony Mezzacca; Bob Brandle, p 131: Tony Mezzacca; Bob Brandle, p 133: LRP photo file; Tom Klimaski, p 135: both Craig Robertson, p 137: Ed Jacobs, LRP photo file; Dave Walker, p 138-139: LRP photo file; Swede Plaut, p 140: Tony Mezzacca, 141: LRP photo file; Randy McKee, p 143: LRP photo file, p 145: Rich Taylor.

CHAPTER 6

p 147: LRP Photo file; Randy McKee, p 148: Phil Rosette, LRP Photo file; Randy McKee, p 149: Henryk Szamota, p 151: LRP photo file; Randy McKee, p 153: Tony Mezzacca; Doug Petersen, LRP photo file, Mac DeMere, p 155: LRP photo file; Bruce Lorry, LRP photo file; Randy McKee, p 156: Jean Constantine, John Vorisek, p 157: Carl Rotter, LRP photo file; Randy McKee, p 159: both George Alderman, p 160-161: LRP photo file; Randy McKee, p 163: LRP photo file, LRP photo file;Geoffrey Hewitt, LRP photo file; Randy McKee, p 165: LRP photo file, Sam Posey.

CHAPTER 7

p 167: Tony Mezzacca, p 168: LRP photo file; Geoffrey Hewitt, p 169: LRP photo file, p 171: LRP photo file; Geoffrey Hewitt, p 172: Jean Constantine, Rich Taylor, p 173: Tony Mezzacca, LRP photo file; Hal Crocker, p 176: LRP photo file; Stan Clinton, LRP photo file; David Betz, Joe Corbett, p 177: Denis Tanney, LRP photo file, p 178-179: LRP photo file; Randy McKee, p 181: LRP photo file, LRP photo file; *IMSA Arrow*, p 183: LRP photo file; Jack Dolan, LRP photo file; Randy McKee, LRP photo file, p 184: JVS Enterprises, JVS Enterprises, Jan Hyde, p 185: LRP photo file, LRP photo file; David Betz, p 186: LRP photo file; David Allio, Ed Jacobs, p 187: Paul Gulash, LRP photo file; Randy McKee, LRP photo file; Randy McKee, LRP photo file, LRP photo file; Randy McKee.

CHAPTER 8

p 189: Ed Jacobs, p 190: Jean Constantine, Rich Taylor, p 191: Joe Corbett, Randy McKee: Ellen Nelson, Joe Corbett, Joe Corbett, p 193: both LRP photo file, p 194-195: LRP photo file, p 197: LRP photo file, David Moore, p 200: Jean Constantine, p 201: LRP photo file; Stan Clinton, Randy McKee, Randy McKee.

CHAPTER 9

p 203: LRP photo file; Fred Rosvold, p 204: AAVR, Inc. Robert F. Williams, LRP photo file; John Firman, p 205: background Jean Constantine, Jean Constantine, LRP photo file; Rodriguez/Raphael, Jean Constantine, p 206: Robert Selkowitz, Bill Hibbell, Ed Hyman, p 207: Denis Tanney, Ed Hyman, Craig Robertson, Carl Rotter, p 208: Denis Tanney, p 209: Rick Kopek, p 210-211: LRP photo file; Randy McKee, p 212: LRP photo file, Jean Constantine, p 213: Jean Constantine, p 214: Jean Constantine, Denis Tanney, p 215: Jean Constantine, Ed Hyman, p 216: Denis Tanney, LRP photo file.

CHAPTER 10

p 219: Mac DeMere, p 220-231: Rich Taylor, p 232: LRP photo file.

Introduction

This book started out to be a simple year-by-year history of Lime Rock Park, one of the oldest, continuously-operated road racing circuits in the United States. But as I started writing it, two factors emerged to alter that naive concept into something far more complex.

One, Lime Rock does not exist in a vacuum, so that changes in American racing—the increasing professionalism of the Sixties, for example, or the founding of IMSA in the Seventies—had a profound effect on day-to-day events at Lime Rock.

Two, Lime Rock did not just react to these changes, Lime Rock directors like John Fitch, Jim Haynes and Jim Shane *caused* many of the changes that have revolutionized American racing over the past thirty-five years. Early on, I made the decision to try and explain the importance of Lime Rock by explaining the expansion of American road racing in general and Lime Rock's special place within that phenomenal growth.

This means I have had to make some difficult choices. You will not read about how I won the 1985 New York Region Championship in ITA, for example, nor how I was awarded the Goodyear Cup at the 1991 BMW Vintage Festival. Nor will you read about similar feats by my amateur racing friends and compatriots.

Tens of thousands of people have raced, crewed, flagged and officiated at Lime Rock over the past 35 years. Some sort of event has taken place an average of 200 days each year. Obviously, I had to make some choices of what I could cover and what I had to leave out. The people, cars and races I selected are representative of thousands of others, many of whom may be equally important. I apologize to anyone who should have been included and was overlooked.

There have been races at Lime Rock that will be remembered forever, because they literally changed the course of American racing. The 1959 Formula Libre event, for instance, because it successfully mixed amateur and professional road racers with professional USAC oval track drivers, starting a dialogue that directly resulted in the format of today's CART IndyCar series.

Or take the 1970 Lime Rock Trans-Am, the most glorious event from the five memorable years of that fondly-recalled series. In a very real sense, the original Trans-Am led to every subsequent "door-slammer" series, from IMSA GT to Kelly American. Trans-Am and IMSA GT made household names of drivers like Mark Donohue, Peter Gregg and Al Holbert.

Or take Lime Rock Little LeMans. In 1957, this introduced the concept of Showroom Stock endurance races,

My friend Rick Pocock celebrates another ITB victory in his Alfa-Romeo GTV, a photo that says so much about not only the past and present of Lime Rock Park, but the pure joy of motor racing, as well.

leading directly—albeit 20 years later—to IMSA Firestone Firehawk and the SCCA Endurance Series.

Lime Rock's *Car and Driver* Challenge races of the mid-Seventies gave immediate legitimacy to Showroom Stock Sedan—then a Regional class in its first tentative year. Today, Showroom Stock is by far the most popular SCCA racing class, credited with saving amateur racing.

The Challenge races led to the concept of racing little sedans for cash prizes. They also sparked the idea of having everyone not just in similar but in *identical* cars...which led not only to the various professional one-design VW/Bilstein Cups and the Renault Challenge, but by extension to the latest ideas in one-design racing...Barber Saab, SCCA Spec Racer and the Shelby Can-Am.

Then there's Henryk Szamota's wonderful NARRC series, which for nearly thirty years has given Northeast Divison Regional racers a season-long Championship headquartered at Lime Rock and fully as competitive as any racing series in the country. There are other Regional series, now, but they are merely pale imitations of NARRC.

What has made Lime Rock Park so special for nearly four decades? A careful balance between the casual atmosphere of a Connecticut country club and the effortless professionalism of a well-run resort. One of the most beautiful natural settings of any race track in the world. A tradition that stretches back to the time when cars were *designed* to slide around corners, when the booming engines were up front, the suspensions were stiff and the frames flexible and the tires no wider than your hand.

By the same token, Lime Rock is as modern as tomorrow, hosting everything except NASCAR and CART, but including IMSA GTP, the fastest and most technically advanced racing cars in this country. While Lime Rock has always respected tradition, it has also always been one of the most forward-looking tracks in the country. As former Lime Rock President Jim Haynes says, "We'd try anything."

But there's something more, something indefinable about Lime Rock. As Winston Churchill said about Russia, "It is a riddle wrapped in a mystery inside an enigma." I think other racers will agree with me that Lime Rock is not an easy track to master. The "Lime Rock Expert" who seems to hold an inordinately large home track advantage is a very real phenomenon.

This is another way of saying that you cannot just waltz in here and go fast, right out of the box. Lime Rock is a race track that demands prolonged study, that can be subtle and treacherous, and yet one that can, on that rare day when you get everything right, provide immense pleasure. If it is possible to love a race track, I know thousands of people, myself included, who *love* Lime Rock.

Racing is a very special hobby; a very special way to make a living. Racing is constantly different, every day, every lap, every corner, so it never becomes dull. It is a constant learning experience, a constant challenge. It demands a surprising amount of physical effort, plus a degree of mental concentration that can leave you exhausted after a half-hour race.

It's also *fun.* Professional racers don't race for the money, though they might say they do. They don't race for the glory, because racing is mostly hard, dirty work, punctuated by moments of sheer panic. People race because racing gives them sensations and satisfactions which nothing else approaches. Do it right, and that's the best day a person can have.

Make it sunny afternoon at Lime Rock Park in May or September, temperatures in the sixties, put yourself at the back of the pack because you missed qualifying, so that now you have to work your way to the front to win on the last lap, and, well...life doesn't get any better than that.

Prologue

Roger Love's stripped-down Willys 77 competing in the Automobile Racing Club of America's Briarcliff Trophy Race of 1934. ARCA had a tremendous influence on American racing—it led directly to the postwar Sports Car Club of America, it organized some of the first European-style road races in this country and the modified American passenger cars that many members raced were the inspiration for not only hundreds of American road racing "specials" in the Thirties, Forties and Fifties, but also influenced stock car racing in the Thirties.

The first organized motor race was run just about a century ago, on July 22, 1894. The fastest vehicle, the Comte de Dion's steam tractor, averaged 11.6 mph over the public highways from Paris to Rouen. This type of open road race continued to be the norm in Europe until 1908, when Grand Prix racing was temporarily banned because it was impossible to control the huge and excited crowds who pushed into the rutted dirt roads in front of primitive cars that even then could easily speed to over 100 mph.

By 1904, there were 100 mph races in America, too—the Vanderbilt Cup on Long Island, the Savannah Grand Prix—that ran over public roads. These were just as dangerous, so in 1908 Willie K. Vanderbilt built the first private toll road exclusively for motorcars—the Long Island Motor Parkway from Queens to Lake Ronkonkoma—then temporarily closed it to traffic in order to run the Vanderbilt Cup races in 1908, 1909 and 1910. Willie K., in other words, ran the first road races in America held on a private, closed course.

There had been closed circuit racing before, of course. Every county fairgrounds in America boasted a half-mile dirt oval used for horse racing. It took no great leap of imagination to substitute the new mechanical steeds for horseflesh. And in 1906, Englishman H.F. Locke King had created the first purpose-built automobile racing track, a gargantuan banked oval on the grounds of the Brooklands estate in Weybridge, Surrey. Three years later, Carl Fisher built the smaller, less radical Indianapolis Motor Speedway, the first full-time automobile racing track in this country. Fisher's original plans called for a road course to wander through the infield of the Indy oval, but the road course was never built.

From about 1910 until after World War II, racing in Europe and America dramatically diverged. With the notable exception of Brooklands, most European racing was over public roads closed to normal traffic for the occasion. There were "round-the-houses" events like Monaco, temporary circuits like LeMans made from a short loop of country roads and over-the-road races such as the Mille Miglia that differed little from Paris-Rouen except that the average speeds were now much greater and the crowd control more stringent.

The Germans were typically more advanced, in that they had already figured out that the best way to go road racing was on a permanent, purpose-built track. The Avus circuit near Berlin was started in 1913 and finished in 1920; the intimidating Nurburgring was begun in 1925 and completed in 1927. With a total length of 17.5 miles—though usually, only the "short" 14.17 mile North Loop was used—the old Nurburgring was and is one of the wonders of motorsports.

Starting well before World War I, Americans no longer raced on the street, at least not officially. They competed on oval tracks, dirt or paved, ranging from the dirt quarter-mile on the outskirts of nearly every town to Indy's 2.5 miles of brick. One exception was an annual road race held over an 8.5 mile circuit laid out on the gravel roads around Elgin, Illinois. From 1910 until 1920, and again for one year in 1933, the Elgin National Road Races were open to Indy cars. Legends like Barney Oldfield, Ralph DePalma, Eddie Rickenbacker and Wilbur Shaw raced at Elgin. In 1933, there was also a class for stock American passenger cars, dominated by Ford V-8s. In a very real sense, you can say that the Elgin road races led not only to professional road racing as we know it today, but to stock car racing, too.

ARCA

In 1933, three wealthy Westchester teenagers, Barron, Jr., Sam, and Miles Collier, incorporated the Automobile Racing Club of America and started holding road races on a .7 mile course over the dirt roads of their father's estate in Pocantico Hills. They named their rudimentary course The Sleepy Hollow Ring, and from 1933 through 1941, they held a series of events for cars like Bugattis, Rileys, Austins, MGs and fastest of all, stripped down Willys 77s.

ARCA also held round-the-houses races at Briarcliff in Westchester county, Marston's Mills on Cape Cod, up Mount Washington in New Hampshire, at Montauk on Long Island, at Alexandria Bay in upstate New York and around the exhibition buildings at the Flushing Meadows World's Fair in 1940.

ARCA was also involved in running a revived "Vanderbilt Cup" at Roosevelt Field on Long Island—the airport from which Lindbergh left for Paris—in 1936. They fashioned a 3.98-mile circuit by connecting runways and access roads. The organizers posted a huge purse, enough to attract not only many American oval track professionals, but also European stars like winner Tazio Nuvolari and Jean-Pierre Wimille. In 1937, virtually every Grand Prix team showed up, including the all-conquering Mercedes-Benz and Auto Unions. Bernd Rosemeyer won the 1937 Vanderbilt Cup for Auto Union; Dick Seaman was second for Mercedes.

Remembering it twenty-five years later, eyewitness Ralph Stein wrote, "The Alfas of the previous year had shocked us with their power...but the Germans of 1937 overwhelmed us, turned our knees to jelly, and stunned our brains with the earth-shaking crash and scream of their exhausts, their unbelievable speed." If nothing else, the Vanderbilt Cups of 1936 and '37 introduced thousands of Americans to the excitement of world-class road racing.

SCCA

In 1944, a small group of American enthusiasts got together and incorporated a successor to ARCA, which they called the Sports Car Club of America. Many of the members were the same, including the Collier brothers, George Rand, Charles Moran and Briggs Cunningham. After World War II, the SCCA started running European-style road races wherever they could get permission.

Modern day American racing began on October 2, 1948, with the first event at Watkins Glen in central New York. Cornell student Cameron Argetsinger convinced the government of this quaint little town on Lake Seneca to let him put on a race over 6.6 miles of village streets and country roads. Frank Griswold won with a pre-war Alfa Romeo 2900; Briggs Cunningham was second in his old Bu-Merc special that he'd had built for the World's Fair race in 1940. Within a year, there were similar road races all over the country.

Future star John Fitch competing at Watkins Glen in 1949, driving his slightly-modified MG-TC. There was an immense charm to racing in those days, though the dangers are quite evident, too.

There were four types of sports car racing in those early postwar years. People raced sports cars on small oval tracks—Phil Hill started this way—which was a silly thing to do and extremely hard on the bodywork. They raced through the streets of small towns and out into the surrounding countryside and back; Watkins Glen, Bridgehampton on Long Island, Elkhart Lake, Wisconsin, Palm Beach in Florida, Pebble Beach on Monterey Peninsula. They also raced on public airports—the biggest and most famous was in Sebring, Florida.

Best of all, General Curtis LeMay, head of the Strategic Air Command, was a great sports car enthusiast. He encouraged the SCCA to race on his SAC bases from coast to coast. Round-the-houses racing was exciting, but dangerous. Airport racing was much safer, but it was often hot, windy and relatively undemanding for the drivers. It had one major advantage, however. An airport could be easily closed off to civilians and the spectators kept safely away from the speeding cars.

By 1952, there were Cunninghams and Allards capable of 150 mph. It

Jaguar XK-120s racing through the streets of Bridgehampton in 1950. Back then, "race prep" consisted of removing the hubcaps and taping the headlights; safety equipment was a polo helmet. Spectators could literally reach out and touch the cars roaring past.

was obvious even then that the round-the-houses racers were courting disaster. It came during the second lap of the 1952 Watkins Glen Grand Prix. Briggs Cunningham, John Fitch and Phil Walters were right behind and/or next to Fred Wacker's Allard J2 as they roared into Old Corning Hill, fighting for the lead. Wacker bounced his rear wheels off the curb and charged after Cunningham's C4R which had passed him under braking. Behind him, Wacker left a dozen injured spectators and a dead seven-year-old boy who had been sitting on the curb. The race was stopped.

The next spring at Bridgehampton, Bob Wilder was killed practicing in Jack MacGregor's J2X Allard, then the race was stopped when Harry Gray crashed his C-type Jaguar, narrowly missing the crowd. These fatal accidents at Watkins Glen and Bridgehampton effectively stopped round-the-houses road racing in America. The Watkins Glen races were moved out of town to a safer circuit made up of country roads; the Bridgehampton road races were temporarily abandoned.

Permanent Circuits

The obvious—but expensive—solution was to build permanent road courses where the spectators could be kept separate from the hurtling cars and the drivers could be given enough run-out room to avoid calamity after some minor mistake. Because of the expense, most road racing circuits remained nothing more than gleams in the eyes of promoters.

Then came the spring of 1955. On May 26, Formula One World Champion Alberto Ascari was killed at Monza. On May 30, Indy champion Bill Vukovich was killed during the Indianapolis 500. Less than two weeks later, on June 11, 1955, came the famous 24 Hours of LeMans.

Mercedes-Benz team manager Alfred Neubauer had paired Juan Manuel Fangio and Stirling Moss, Karl Kling and Andre Simon, John Fitch and Pierre Bouillion-Levegh in fabulous 300 SLRs. Levegh was running fifth overall when Mike Hawthorn's Jaguar passed Lance Macklin's Healey then braked to come into the pits. Macklin swerved to avoid

John Fitch's teammate, Pierre Levegh, crashed their Mercedes 300 SLR in the crowd at LeMans in 1955. The worst accident in racing history, this disaster led directly to safety improvements at existing tracks and the building of dozens of new, safer race circuits all over the world—including Lime Rock Park.

him, Levegh swerved to avoid the Healey, hit the bank and catapulted through the crowd of spectators, his magnesium-bodied sports/racer exploding like burning shrapnel.

Not just Levegh, but at least 81 spectators died as well. John Fitch spent the next seven hours trying to convince Herr Neubauer that Mercedes should withdraw as a sign of respect. Fitch was right and the Daimler-Benz Board of Directors finally ordered the cars withdrawn sometime after midnight. Fangio/Moss were leading by two laps at the time.

The calamity at LeMans, coming so soon after the deaths of Ascari and Vukovich, brought a dramatic reaction from all over the world. Switzerland banned racing, the Grand Prix season was shortened, many races were cancelled. Mercedes withdrew from racing at the end of the season, Lancia sold their Formula One cars to Ferrari, John Fitch's Cunningham teammate Phil Walters retired and Fitch himself decided to come back to America. Back home, the AAA—which had been sanctioning competition since before World War I—soon decided that racing was bad for their image. Tony Hulman repaved his Indianapolis oval to make it safer and founded USAC to sanction professional racing in opposition to the AAA, which soon abandoned racing altogether.

Most significantly for our story, people all over America who had been thinking about building a permanent road racing circuit decided that now was the time. Laguna Seca, Watkins Glen, Bridgehampton and Lime Rock, among others, were started in the aftermath of the LeMans tragedy. It was a dramatic turning point in the history of motorsport; from now on, road racing on permanent, purpose-built circuits would be the rule. The Golden Era of American road racing was about to begin.

1.
The Early Days

Lime Rock Park Opening Races, April 28, 1957: Gordon MacKenzie's Jaguar XK-120M leads Tony LaPalme's Mercedes 300 SL up the hill in the C-production race. After 10 laps, La Palme was second and MacKenzie third behind Bob Rubin's similar Jaguar. They turned hot laps of 1:20. Walt Hansgen's D-Jag was the fastest car there that first weekend. He zipped around in 1:10.4 while winning C-modified. This became the first outright track record.

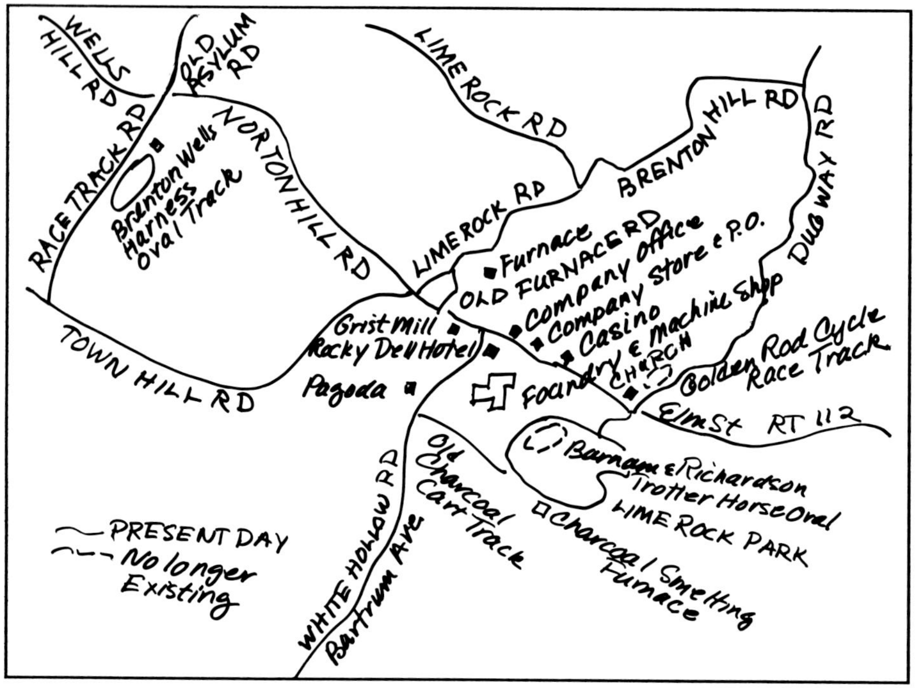

Map of the Lime Rock area shows where the early iron industry buildings and nineteenth century race tracks were in respect to Lime Rock Park. It's obvious that Lime Rock has never been a quiet little town in Connecticut; it's always been a commercial center.

Lime Rock is a tiny hamlet in the Berkshire foothills, located in the center of what is commonly called the "Northwest Corner" of Connecticut. This is lovely, pastoral countryside, closely bordered by the hills of Massachusetts on the north and the farms of New York on the west.

Today it is almost disturbingly quiet, but for two centuries Lime Rock was a continuously bustling center of heavy industry...noisy, dirty and gray with the smoke of fires that burned 24 hours a day, seven days a week and made the streams run black. It was not a particularly picturesque spot to spend a lost weekend.

The very first white settler, Englishman Thomas Lamb, came to Lime Rock in 1730 because he'd heard rumors about a deposit of iron ore so rich it literally stained the ground brown. Lamb built a forge and a foundry, a grist mill and a sawmill. By 1734, he and his workers were shipping maleable iron all over New England. The woods echoed with the clank of the forge, the ring of the hammers and the shriek of the sawblades, not to mention the clouds of spark and charcoal smoke from the iron ore ovens.

During the Revolutionary War, Lime Rock iron was made into muskets, cannon, anchors, chain and other munitions. The sister ships U.S.S. *Constitution* and *Constellation* were fitted out with Lime Rock iron.

Half-a-century later, Lime Rock was still a thriving industrial center. The Salisbury Iron Company built its first blast furnace in the early 1830s at the corner of Lime Rock Road and Furnace Road, with a waterwheel in the Salmon Kill to work the huge bellows. Charcoal was made on Sharon Mountain and carried down to the the furnace on a wagon track that today forms the infield entrance road to Lime Rock Park.

The thing about a blast furnace is that it is kept running continuously. At the north end of Lime Rock was Hoppertown, where the iron ore was poured downhill into the blast furnace. What with the day and night wheezing of the bellows, the acrid pollution from the furnace and the glow of the molten iron, a typical evening at Lime Rock in the early-nineteenth century must have been noisier and smokier than any IMSA race ever dreamt of being.

By the 1840s, the Lime Rock Ironworks was making wheels for railroad cars and even had its own railroad stop, Lime Rock Station on the Housatonic Railroad. In 1864, local businessmen William Henry Barnum and Leonard Richardson bought the iron company, built a new, even larger blast furnace and expanded until there were factory buildings lining what is now Route 112 from Well's Hill Road almost all the way to the Lime Rock Park office.

The workers lived in and around this Piranesian nightmare in a "model" company town that included not only workers housing but a school, a combined clubhouse/theatre/ballroom called the Casino and the Rocky Dell Hotel, a boarding house for unmarried workers that had its own dance hall on the third floor. The workers lived on the south side of town, the Barnum and Richardson families lived on the north side of 112 in elaborate Victorian homes that are still the finest in Lime Rock. Indeed, there is very little Colonial building in Lime Rock as there is in most of the surrounding towns; almost everything dates from the mid-nineteenth century and was built by Barnum and Richardson.

The First Race Track

According to Reed Williams' *Memories of Lime Rock*, the first race track on the Lime Rock Park property was an oval where the wealthy Barnums and Richardsons trained their trotters in the 1870s. It was located about where the Skip Barber School skidpad is now. Explains Williams, "Racing fever gripped Lime Rock. Horses and mules on the public highway were raced against others going in the same direction." This may have been the origin of Friday Night Street Racing. No doubt there was betting, for betting on horses was a nineteenth century avocation that cut across all social and economic boundaries. Life in Lime Rock in those years must have had all the earthy *joie de vivre* of a Mo-Hud Regional.

Reed Williams claims that lovely Trinity Church (across the street from Lime Rock Park)—designed by the famous New York City architect Richard Upjohn and consecrated in 1874—actually came about because the Barnum and Richardson men were racing their carriages on the way home from church in Salisbury on a rainy, windy day. Their cold, muddy wives united to force the wealthy sportsmen to invest in a new church, closer to home. Some local historians claim that this is the origin of the bumper sticker "She said it was either me or racing, and Lord, I'm gonna miss her."

"In the 1880s," says the erudite Williams, "a new racing fever hit Lime Rock—bicycle racing." The Golden Rod Cycle Race Track was constructed between Dug Way and the rear of Trinity Church and a regular race series was started with cyclists coming from all over New England to race at Lime Rock. Racers from the local club also competed at other tracks for a season-long regional championship not all that different from the NARRC series based at Lime Rock today.

Near the end of the century, Brenton Wells built a harness racing oval track on Race Track Road that drew its own crowds. The bicycle racing oval track closed around the turn of the century, however, and the small grandstand at Mr. Wells' track burned in 1915. This

(Top) Trotting races were all the rage in Lime Rock at the end of the nineteenth century. There were two oval tracks in town, plus a bicycle track.
(Below left) Huge industrial buildings lined the Salmon Kill in Lime Rock for over a century until washed away in the 1955 flood.
(Below right) The Rocky Dell Hotel and downtown Lime Rock as it looked in 1930.

(Right) Sunday morning, 1954: Jack Fisher and Jim Vaill in Fisher's MG-TC, kicking up a cloud of dust in the soon-to-be-replanted sweet corn fields of Jim's father's Lime Rock farm. They're racing through what is now the C parking area, in the Lime Rock Park infield.

big oval track remained derelict for almost four decades. Then Dr. Robert Noble restored the old harness track and held races on both Saturdays and Sundays up into the Seventies. You can still see the track and some of its buildings on Race Track Road.

The local iron industry played itself out before World War I and most of the company town was sold to individuals in the Twenties by real estate investor Alfred Stone. In 1928, the old mill was turned into a paper-making factory by Dard Hunter, but that venture went bankrupt in the Thirties.

Lime Rock Park

By the early-Fifties when our story begins, the old Barnum and Richardson factory buildings were standing empty along the Salmon Kill; the old smelting furnace was a silent stone sentinel in the backyard of a private home. About the only heavy industry left in Lime Rock was a sand and gravel company operated by Jim Vaill. It wasn't as noisy as a blast furnace nor as pugnacious as a bicycle racing crowd, but the sprawling gravel pit wasn't exactly an asset to the neighborhood, either.

Jim's father, Frank Vaill, was a prosperous local farmer with extensive property on Wells Hill Road as well as 385 acres in Lime Rock. Frank grew sweet corn and potatoes. It was Jim's idea to start the sand and gravel business as a way to supplement the seasonal income from the farm.

Two of Jim's friends were ski resort owner Jack Fisher and a young car enthusiast named John Clark. Fisher had an MG-TC, and he and Clark would bring it down to Jim's dusty gravel pit on Sunday afternoons and roar around, pretending they were racing drivers. Their unpaved loop was right about where the infield C parking area is now.

As Jim Vaill tells the story, "One Sunday in the spring of 1955 we were sitting there eating onion sandwiches and drinking beer, and these people from the SCCA showed up to find out about our race track! It had never entered my mind before then, but I decided maybe that was a good idea."

According to Gaston Andrey, those "people from the SCCA" were some of the most famous American racing drivers of the time, including Briggs Cunningham, Sebring winner Bill Lloyd, Fred Procter and himself. "We went up to Lime Rock," he says, "and we walked around this gravel pit with Jim Vaill and helped him plan where a race track might go. We spent a lot of time down there looking at things."

Encouraged, Vaill went to a Lime Rock town meeting and showed a movie he'd borrowed from the SCCA about LeMans. He told his neighbors he wanted to build a similar track and hold similar events. The town agreed to go along, voting by a two-to-one majority to let Jim build a track. Frank Vaill also agreed to let Jim use his land, and even put up the funds to get the project started.

"I think they all imagined hundreds of thousands of spectators buying tickets and meals and hotel rooms and making them rich," says former track director John Fitch. Of course, it didn't work out that way. As John once said about his Corvette team's finish at Sebring, "It was not as good as we had hoped, but better than we deserved." The same could probably be said of this underfinanced, surprisingly casual decision to build a race track in the midst of sweet corn fields surrounding a gravel pit in the depths of the Connecticut countryside.

In the early summer of 1955, Jim Vaill got an old aerial photo of his father's property and using white paint, drew a race track on the photo. "You could tell just by looking at the aerial photo where the track would have to go," explains Jim. "You could see where the sinkholes were, where you could or couldn't run the track. The Lime Rock track was dictated by the terrain."

Once he'd drawn it out, Jim simply got on his Caterpillar D-8 and started cutting a swath through the trees. "I just started off through the woods. At first, I had to follow the brook. The land was so thick with enormous elms and maples that you couldn't get through, even with a bulldozer. I got down somewhere around where Big Bend is, and the woods were so thick, I couldn't see to find my way back."

The Salmon Kill in those days meandered through the edge of the gravel pit in a series of loops and oxbows. Jim straightened it out and moved it from what would now be the main paddock to the opposite side of what is now the front straight. "I just took the dirt I had to move to make the track and dumped it into the stream to fill it up and change its course," Jim says matter-of-factly. Environmental laws wouldn't let you even think about that, today.

In August of 1955, there was a hurricane and the Salmon Kill overflowed, washing away the old mill

(Continued on page 15)

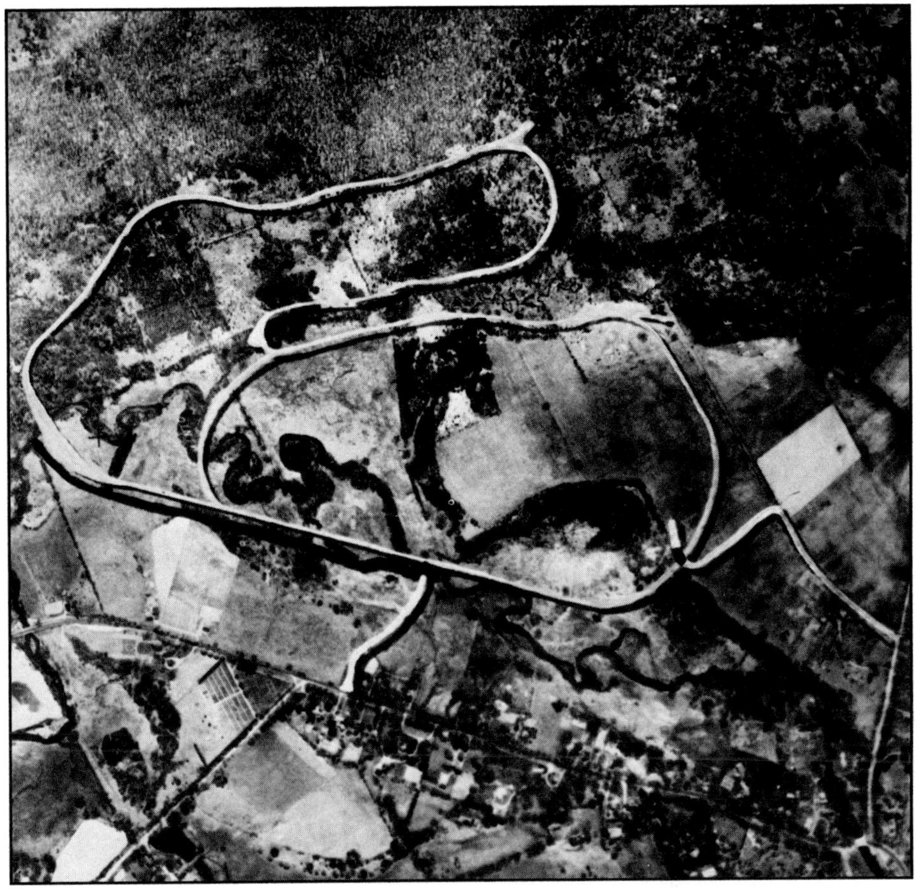

Jim Vaill's original 1955 drawing for Lime Rock Park shows the general layout, the Mountain Circuit and even the infield and outfield entrances. The Salmon Kill meandered through what is now the A paddock.

JIM VAILL

Jim Vaill is not a famous racing driver. "I drove my street car around Lime Rock once at a 1:18," he says. "That's the nearest I ever got to racing." But he has had more of an effect on American racing than many people who are household names. Jim—almost single-handedly in the beginning—conceived, created and ran Lime Rock Park.

But the race track at Lime Rock was only a short interval in his life. Jim Vaill is one of those people who can get things done, without help. He just does it. Before building the track, he was a farmer running one of the largest commercial potato and sweet corn operations in the Northwest Corner. "I grew up on the farm," he says. "As a child, I can remember my mother doing a lot of canning in the summer. We grew our own vegetables, our own pigs, chickens and eggs, milk, butter, beef. We cured our own meat and made sausage, bacon and ham.

"We were almost self-sufficient. In the fall, my father would buy a barrel of flour, a barrel of sugar, a barrel of salt and some cartons of salted codfish. And Uneeda Biscuits. Other than that, we made everything ourselves."

This early training in self-sufficiency set the pattern for the rest of Jim's life. In addition to farming, he ran his own sand and gravel business from the Lime Rock site, doing most of the work himself. This was before, during and after the construction of the race track. "We started the gravel business in 1950, and we were still taking gravel out of there for a year or more after the track opened," he says.

Talk about self-sufficient; Jim personally planned the track, bulldozed the site, put together the financing and ran not only the business, but the races themselves. This covered the period from his first idea to build a track in 1955 to the sale of the track in 1963.

After that, he bought and ran the Iron Masters Motel in Salisbury, then did quite a bit of real estate development and residential building. Today, he's retired, splitting his time between houses in Antigua and Lakeville. His Lakeville home is a modern house built on the property of the dairy farm where he grew up. In that sense, he has spent literally his entire life on the same piece of land.

But Jim Vaill is no simple country boy. He is a shrewd businessman and also the nicest man you'd ever want to meet. He loved racing, in the way that only an enthusiast who does not race can love racing.

Think of it. With no experience and no money, in 1956 Jim spent the whole year working full-time building his race track, raising money and convincing his friends and neighbors to not only let him pursue this folly, but to help bankroll it. Jim Vaill wasn't a starry-eyed kid at the time, either. He was 38 years old when he built the track. Why do this thing? "Well," he says. "Racing was fun." ■

(Above) Opening Day, April 27, 1957: Lime Rock builder Jim Vaill in front of the new tower. (Opposite page) The track under construction in the summer of 1956.

WALLACE
15
Firestone
16
7

buildings and the bridge. It formed a lake in The Hollow and Jim's gravel pit was under 10 feet of water. As he says, "The D-8, the D-7, the D-4, six dump trucks, the whole screening plant, the crushing plant, the shop—it was all wiped out by that flood."

The Lime Rock Corporation

It took six months before Jim Vaill was able to start working on his track again. In the meantime, he raised $100,000 by selling stock in his new The Lime Rock Corporation. "I was hoping for some big investors," says Jim. "I went into New York City a number of times, trying to get some of these wealthy big shots to finance me. But the most I ever got from any one person was $5000. Nobody gave me more than $5000. Most of the investors in Lime Rock Park were just small, local businessmen who gave me $1000 or $2000 because they thought the track would be good for their business. There were over forty shareholders."

One twist that was to create trouble later on was the fact that Jim's father, Frank Vaill, owned the land upon which the track was built. The Lime Rock Corporation invested all that time and money in building the track, but they never owned the property. This was one of the few mistakes Jim Vaill made. It seemed like a good idea at the time.

"My father wanted to sell us the property for $50,000," he says. "But neither I nor the corporation had enough money to buy it. Leasing seemed like the only way we could afford to get the track built for the small amount of money we had to spend."

A smarter thing Jim did in the fall of 1955 was to call John Fitch. Now when Mercedes-Benz retired from racing at the end of the 1955 season, it left Fitch out of a ride. He and his family returned from Switzerland to their home in Stamford, Connecticut. As John says, "I was looking for something new to do, and one of the first phone calls I got was from a group of sports car enthusiasts who were building a race track up in the woods by Lime Rock."

Through sheer coincidence, John Fitch had been behind Freddie Wacker's Allard when it hit the curb at Watkins Glen in 1952, and he had been drinking coffee with Pierre Levegh's wife, awaiting his turn in the 300 SLR, when Levegh crashed at LeMans in 1955. Fitch himself had also crashed heavily at Rheims in 1953, destroying Briggs Cunningham's C5R. More than most people, John had firsthand knowledge of the appalling hazards of racing on public

(Opposite page) Industrial designer Raymond Loewy had his New York office prepare dozens of color sketches for proposed facilities at Lime Rock. Most of the sketches have been lost over the years. This one shows pits similar to those at LeMans, with permanent timing and scoring booths and spectator bleachers above the enclosed areas for mechanics.
(Below) Loewy's futuristic pedestrian bridge would have connected infield and outfield.

Among other projects, E. Farnum Spicer was John Fitch's partner in John Fitch Motors and also a director of Lime Rock. This is one of his shares of stock in The Lime Rock Corporation, dated December 18, 1958.

people in his office, and they had been responsible for the 1953 Studebaker. That was a lovely car."

In 1956, when he heard that his friend John Fitch was involved with the exciting project of building a new race track from scratch, Raymond Loewy volunteered to design the buildings and landscaping, transforming Lime Rock into what Loewy called "an automotive showplace."

Loewy's exquisite drawings show parklike grounds housing a clubhouse, a museum/exhibition building, restaurants and a motel, timing and judging stands, electronic scoreboards, elaborate pits with spectator stands above them, dramatic bridges between the infield and outfield, a helicopter landing pad and even futuristic mini-trains to carry spectators around the infield. Loewy planned paved pit and paddock areas, enclosed garages similar to those at Nurburgring and parking for 30,000 cars. The facilities were planned to accomodate 100,000 spectators without crowding.

Grand Prix Circuit

In addition to what evolved into the familiar 1.53 mile Lime Rock circuit, Jim Vaill's original drawing also shows an additional "Mountain Circuit" that would extend from the end of the front straight, wander through the woods on the side of the hills for 1.5 miles and rejoin the Lime Rock circuit at The Hook. Aptly named, the exciting Mountain Circuit would have a total elevation change of over 200 feet. Together, the Lime Rock Circuit and Mountain Circuit would form the 3.0 mile "Grand Prix Circuit."

John Fitch immediately seized on this plan. His intent was to bring European professional racing to America. The Grand Prix racers were his friends, and he obviously saw a wonderful opportunity to hold the inaugural U.S. Grand Prix at Lime Rock.

It would be the first time since the Vanderbilt Cup of 1937 that the world's finest cars and drivers would race in this country. To lure them here would require more than a new 1.53 mile road course, no matter how scientifically-designed or interesting. Never ones to think small, both John Fitch and Jim Vaill were desperately anxious for a "full-size" Grand Prix circuit.

In addition, Loewy and Fitch planned a .5 mile "Training Circuit" on the outside of the Lime Rock Circuit esses. The Training Circuit was to have a skidpad circle plus a tight little course that could be used for driver's schools, chassis set-up or vehicle dynamic studies. Loewy even suggested that it be paved in "architectural glass block," to be used for studying skid characteristics on a slippery surface. Like the wet skidpad used by today's racing schools, the glass course would approximate the feel of a four-wheel drift at much slower speeds.

The Real World

Virtually nothing from Raymond Loewy's extravagant plan for Lime Rock Park came to fruition. Nor did the Mountain Circuit, the Grand Prix Circuit or the Training Circuit. Jim

Vaill did cut the route of the "Mountain Circuit" through the woods—thirty-five years later, you can still see the gap in the trees—but The Lime Rock Corporation simply did not have the $500,000 to $750,000 that Loewy figured his scheme would cost. That was a lot of money in 1956.

Jim Vaill personally did most of the construction work on the track with the help of his friend Don Silvernale. It took them most of 1956, working full-time. As Jim says, "My father supported me the whole time I was working on the track."

The 1.53 mile course was bulldozed and graded, then paved with 24 inches of bank run gravel topped by 6 inches of processed gravel and 1.5 inches of bituminous concrete. The gravel came from the site, and

(Continued on page 22)

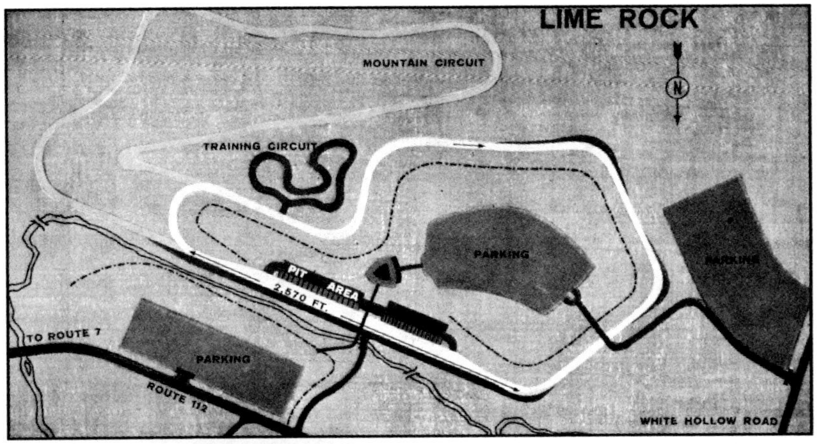

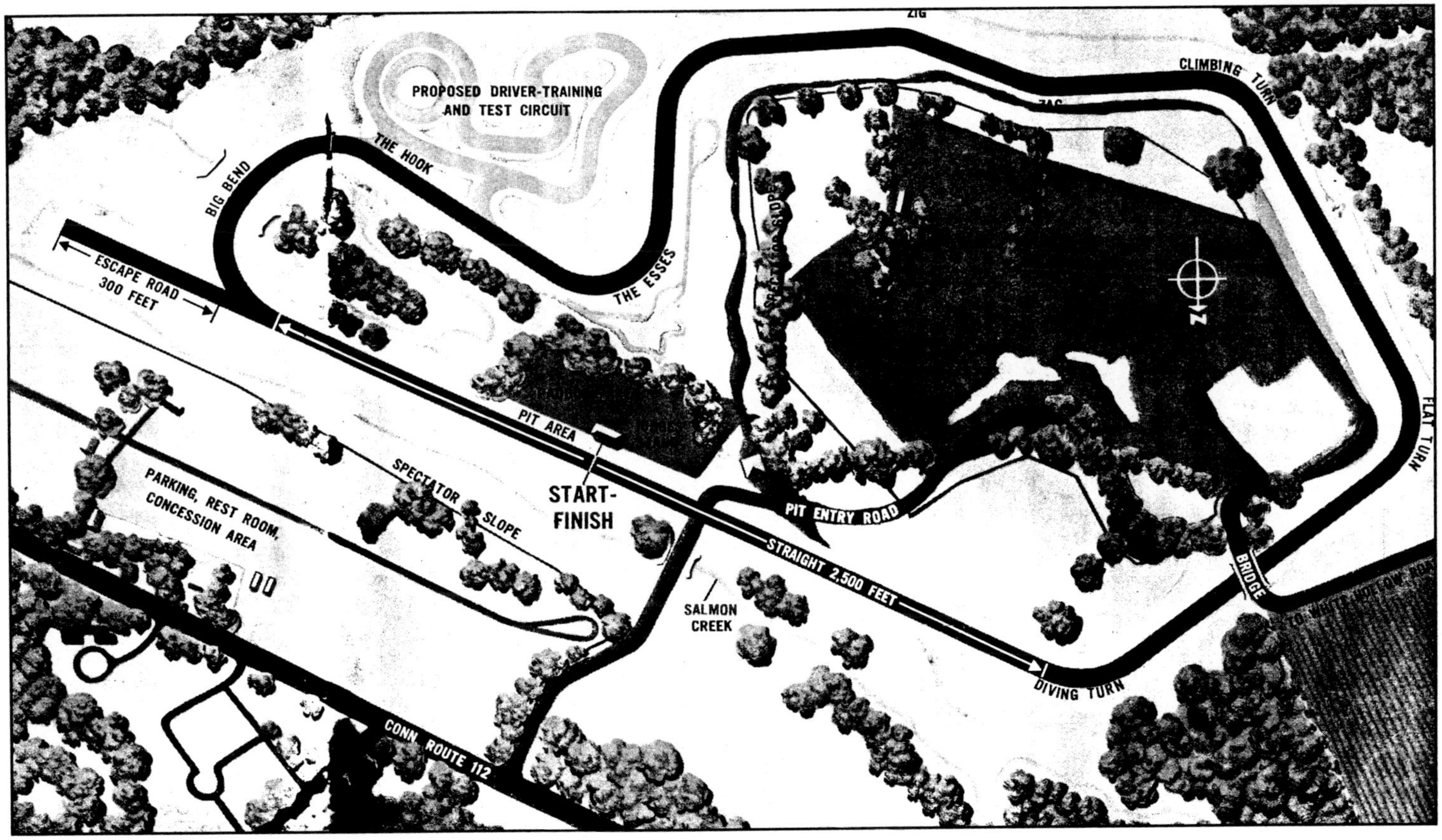

(Above) This map appeared in Sports Illustrated *in May, 1957. You can see that The Lime Rock Corporation was still planning to build the Mountain Circuit and Training Circuit, along with Raymond Loewy's elaborate pits and bridge. The triangular building in the infield was to be a combination tower, restaurant and spectator viewing center. (Left) Another 1957 map shows Lime Rock Park as built. Notice the trees in the infield paddock and the small start-finish "judges' stand." This map is still surprisingly accurate today.*

JOHN FITCH

John Cooper Fitch is quiet and soft-spoken, the very embodiment of the word Gentleman. In his conservative tweeds and driving cap, pipe turned up against the wind, still reed-thin and fit at 75, he might be an Ivy League sociology professor or the president of a small local bank.

Don't be deceived. In the course of an incredibly varied career, John Fitch has been a highly-decorated World War II flier, a test pilot for the government—"We'd take them up, hold the throttle open and see what would happen"—and one of the first sports car dealers in America. He designed, built and marketed the Fitch Sprint and Phoenix automobiles, as well as the famous Fitch-Whitmore Special. He holds dozens of patents, for inventions ranging from a remote-operated tailgun for the A-20 attack bomber to his Fitch Barriers, those yellow barrels that line the highway and which have saved tens of thousands of lives since 1970.

He designed the fictional Burano racing cars and did most of the stunt driving in the Kirk Douglas film *The Racers*, has been a consultant to General Motors on and off for decades and team manager for their Corvette racing programs in the late-Fifties. Today he is involved in high-tech companies dealing with hybrid cooling systems, liquid-cooled brakes and zero-emissions hydrogen fuel cells. He's also a director of the firm that builds the Consulier sports car.

And of course, from 1955 till 1964, he was a director of Lime Rock Park, and in a very real sense, the sparkplug that made that perpetually underfinanced project succeed. Without his knowledge and connections—not to mention, his international reputation—Lime Rock Park would probably never have gotten off the ground.

John Fitch is best known as a racing driver, though that is just one facet of his remarkable life. He was the first American driver to live and race in Europe after World War II and the first to drive for a European factory team. He was the first American to win a major European event since Jimmy Murphy swept the French Grand Prix in 1921 and also the first American to win a Coupe des Alpes in the famous Alpine Rally.

John drove on factory teams for Porsche, Sunbeam-Talbot and Corvette, but mostly for Briggs Cunningham. He also raced a dozen private-owner cars, from a Crosley-bodied special he built himself to Stirling Moss's personal Maserati 250F.

The high-point of John's racing career was the 1955 season, when he

was a full-time member of the Mercedes-Benz factory team. Mercedes fielded the most formidable racing team in history, including Juan Manuel Fangio, Stirling Moss, Karl Kling, Peter Collins, Piero Taruffi, Andre Simon, Pierre Levegh, Wolfgang von Trips, Desmond Titterington and John Fitch. When Mercedes-Benz triumphantly retired from competition at the end of the season, they had won the 1955 championships for Formula I, sports cars and Gran Turismo, the first time that one team had ever made a clean sweep of professional racing.

John added to that total by winning the Irish TT with his partner Stirling Moss and also taking the GT class at the Mille Miglia. At the Mille Miglia, Fitch had only a stock 300 SL gullwing coupe; his navigator was a young German magazine editor named Kurt Gesell who had never before seen a racing car. As John wrote at the time, "Compared to the 300 SLRs of Moss and Fangio, our coupe was as tame as a tractor."

Almost miraculously, after 1000 miles Fitch/Gesell were fifth overall behind Moss/Jenkinson, Fangio, Maglioli's Ferrari and Giardini's Maserati. Of course, they won the Grand Touring class, breaking the class record by more than an hour.

But more incredible, John had actually gone faster than Alberto Ascari's Mille Miglia-winning speed the year before at the wheel of a full-race Lancia D-50. John's Mille Miglia drive is not only the best race of his career but one of the most phenomenal performances by anyone, ever.

Fitch's racing started in June of 1949 and though he ran his last race at Sebring in 1966, he had effectively retired after 1959. Racing was just an interlude for John, one decade out of seven that he has filled with staggering variety and enviable energy. More than a racing driver, John Fitch is an eccentric futurist, an inventor, a bit of a tinkerer, an inveterate entrepreneur. "No," he says modestly. "Mostly I'm an eternal optimist." Exactly. ■

The many facets of John Fitch: (Opposite page, top) after winning the 1952 Watkins Glen GP in a Cunningham; (Opposite page, bottom) after having a Lotus 11 break at Lime Rock, 1958; (Above, left) with the first Fitch Barriers, 1970; (Above, top to bottom) in his 2.0-liter Maserati, Cooper Monaco and Genie, all at Lime Rock.

Team Lotus owned three Mark 11s, shared by six drivers. Among them were Peter Baumberger, Walter Cronkite, John Porter and Walter Rolf. This is one of the team cars, ready to be unloaded for practice at the first Lime Rock driver's school on April 20, 1957. Notice their homemade trailer with 2x8 ramps and Chrysler station wagon tow car.

the paving was done by a local contractor after Vaill and Silvernale had prepared the surface.

When the crew was finished, Lime Rock was paved 28 feet wide on the back straight and 40 feet wide on the front straight. The corners varied from 32 feet to 50 feet wide, with a normal banking of 1 foot, 10 inches. It seems like more, but the elevation change is only 32 feet going up The Uphill and about twice that coming down The Downhill. The Front Straight actually slopes slightly towards the east, so that the lowest point on the course is the apex of Big Bend.

While Jim Vaill was working on the actual course, John Fitch was trying to persuade racers to come

there and compete. "We had such problems," he says. "Such problems. We had monumental battles with the SCCA. I wanted to release a list of drivers to the newspapers, a list of who was going to be in our races. The SCCA said, 'That's against our policy.' It was against their policy to publicize their races!

"I wanted to get name drivers, professional drivers to draw the crowd to our first race. The SCCA threatened to ban everybody who ran in such a race, while the AAA, which was near the end of its involvement with racing, threatened to suspend the licenses of any professional drivers who came. I had tremendous battles with the AAA.

"I even had problems with the volunteer corner workers. They said, 'This is a for-profit operation, and we want to be paid.' Unfortunately, we were not making a profit. The local businesses, the hotels and restaurants and antique shops, thought a race track was a great idea. But even before the first race was run, the townspeople in Lime Rock had formed a protective committee to complain about our racing.

"I think they expected Hell's Angels in leather jackets to disassemble their town. They had no idea at all of what kind of people are interested in road racing. Eventually, the obvious benefits to the town, to the businesses and churches, brought about a whole different view. But in the beginning, there was a lot of trouble.

"Lime Rock never had any money. Even before the first race, I arranged for rain insurance, because if we'd ever been rained out, we would have gone bankrupt. We had no cash reserves at all. I actually had to get a rain gauge and calculate the rainfall probability so I could figure out how much insurance I could afford. It was nerve-wracking."

The First Event

It took Jim Vaill, John Fitch, General Manager John Clark and the other enthusiastic stockholders of the Lime Rock Corporation about eighteen months to get themselves ready to race. When they were finished, they valued the completed Lime Rock circuit at $250,000. Their goal was to be open in time for the 1957 season, and as it turned out, the first racing in the Northeast that year was at Lime Rock. It was so early in the spring the leaves weren't yet fully open on the trees and there was still a serious threat of bad weather.

The first event at Lime Rock Park was not a race but an SCCA driver's school, held on Saturday, April 20, 1957. Some 152 cars showed up, swamping the instructors who'd expected only a handful of students. George Rand—who had been at this road racing business since the ARCA days of the early-Thirties when he ran a Grand Prix Maserati—Fred Procter, George Constantine, Don Vitale, Gordon MacKenzie and John Meyer were the instructors.

In addition, another 121 people showed up for a flagger's school conducted by Gaston Andrey and Don Calhoon. During the school, a Volkwagen Beetle earned the dubious distinction of being the first car to flip in the Lime Rock Downhill; a Triumph TR-2 roadster was the first to flip going into Big Bend. Fortunately, the Triumph's driver had in-

(Above) Lime Rock management on Opening Day: left to right, General Manager John Clark, President Jim Vaill and Director John Fitch.
(Left) The standing start of the main event at Lime Rock on Opening Day. Starter Britt Ragsdale is standing next to the track with SCCA executive Ted Andrews. Announcers Art Peck and Michael-Wynn Wilson are on the roof; John Fitch and Walt Hansgen are in the Cunningham D-Jags.

stalled a rollbar and seatbelts—controversial additions at the time—and so drove home with only a few dents in the bodywork to prove his adventure.

Opening Day

Few people remember that the driver's school absolutely tore up the brand new surface of Lime Rock, leaving it full of potholes. In something of a panic, Jim Vaill got local contractor Sid Terhune to completely repave the track in time for Opening Day.

"We originally put down a macadam surface," explains Jim. "And it hadn't had time to cure. After that driver's school, the surface was just gone. Sid repaved that entire track in asphalt in just three days. I had to hire every dump truck in the area; we had two pavers and two rollers brought down from Sheffield and we worked dawn to dusk. But it was worth it. The new asphalt held up for years. I think Sid charged me $17,000 for the whole job."

In photos from Opening Day, you can see that the track surface is significantly higher than the run-out room around it. The property had been graded to the original track level, but there wasn't time to regrade it to the newly-paved surface. Drivers were warned not to leave the black part, because of the sharp and dangerous drop-off that could flip a car.

The first Registration and Tech started at 9:30 am on Saturday, April

(Above) John Meyer's famous Meyer Special married a big V-8 and a Jaguar frame, all covered over by aluminum bodywork. It was one of the most successful big Modifieds of the mid-Fifties.
(Right) The paddock on Saturday morning, April 27, 1957. Team Lotus is in the foreground, with Peter Baumberger's wife Alix—a famous fashion model of the time—walking across to her husband's Lotus 11. Next to it is Gordon MacKenzie's XK-120M. The photo was taken from the top of Jim Vaill's equipment shed.

27, followed by mandatory practice by race group, starting at 10:00 am. The track closed at 6:00 pm. Sunday's racing started at 2:00 pm, thanks to Connecticut's "blue law."

Race Chairman John Clapp, Jr. included this sage advice in his letter to entrants: "As the track entrance is directly across the street from a church, a certain amount of discretion must be used in the matter of exhaust noise and spirited driving, especially on Sunday morning. Our conduct at this first event and the impression we make on the local people will undoubtedly set the pattern for our future relations."

Sunday April 28, 1957 was happily clear and warm. Lime Rock's track was paved, and the most dangerous trees near the course had been removed. But much of the course was still surrounded by large mounds of dirt, the paddock was an ankle-busting mix of dirt and stones and the pit lane was only coarse gravel. It would be years, in fact, before the pit lane was paved.

Spectators were separated from the racing surface by wire fences, but the only noticeable safety barrier was a length of Armco that protected the new timing and scoring building from errant race cars. There were still road-grading machines parked in conspicuous spots and Jim Vaill's gravel loading conveyor overshadowed the paddock, about where the Driving Impressions building stands today. It made a great spot from which to get a panoramic view of the racing.

Next to the conveyor was Jim's old

(Above) The program from Lime Rock Opening Day. The cover photo was actually taken at Sebring in 1956, and shows winner Juan Manuel Fangio in the Ferrari 860 Monza leading Walt Hansgen's Cunningham D-Jaguar. They're lapping the factory Arnolt-Bristol of Bill Stewart.
(Left) Another view of Opening Day, shot from the outfield spectator area. That's Bob Holbert in his Porsche 550 RS just entering Big Bend.

A pretty good argument can be made that the Golden Age of American Road Racing was more than thirty years ago, from 1957 until 1963. Even full-blown sports/racers still looked like cars instead of fighter planes, the drivers sat out in the open where spectators could see them sawing away at the wheel and the engines were mostly up front, where God and Henry Ford intended them to be. It was a most glorious time.

"Our Gang," friends and relatives of photographer Judi Chatfield overwhelm Dad's DeSoto to get a better view from the paddock in 1957.

Road racing was very different back in the Fifties; there was no money in it. Racing has always been an expensive sport, but the Sports Car Club of America was not just for amateurs, it was for gentlemen who could afford to go racing out of their own pockets. There wasn't a sponsor decal in sight. There was also no such thing as a hired driver in the Fifties, though gentlemen who owned racing cars might lend their cars to other gentlemen who wished to "have a go."

That this second gentleman might be employed by the car owner's company in some undefined capacity—and that his sole job description might be getting around Lime Rock as quickly as possible—was conveniently overlooked. Even someone like Carroll Shelby, who made a handsome living as a full-time racing driver, was still an unpaid amateur in the eyes of the SCCA.

In 1957, when Lime Rock held its first race weekend, each of the eight race groups was divided into two sections, "owner driven" and "non-owner driven." There was a separate prize for each section and the owners were not expected to be as skillful as those who "volunteered" to drive other people's cars.

The Amateur Ideal

This whole crazy attitude had started in England in the nineteenth century, when the same divisions had been put up to keep professional jockeys from competing directly with gentlemen amateurs in horsey pursuits like steeplechase racing and polo. Gentlemen did not compete for money. They raced for excitement, for the satisfaction of doing well, they raced *pour le sport*. This was true even at the highest levels of road racing.

"I was never paid anything by Briggs Cunningham," says John Fitch. "It was all very informal. Briggs was the very ideal of the amateur sportsman. To Briggs, we were all gentlemen racing for the fun of it so there was no need for contracts or for money to ever change hands."

This ideal of amateurism was imported from England along with the first MG-TCs, and permeated all road racing in America at least until the mid-Sixties. Roger Penske applied the first sponsor decal to a road racing car on the East Coast when he put "Telar Special" on the side of his Maserati in 1961. This was considered by old-timers to be in extremely poor taste, if not downright illegal.

A weekend of racing at Lime Rock was not a cut and thrust battle of unlimited budgets, it was a picnic in a park with friends, during which at some time in the day you would get out and drive around the track, depending more on reflexes and bravery than experience and technology. When finished driving, you headed for the bar at the Lime Rock Lodge to recount your adventures to your friends and in turn, listen to their adventures. Racing was a pastime, not a career.

In those days, you could go racing with nothing more than a driver's license, an SCCA membership card and a one-day SCCA school to make sure you knew what the flags meant. You didn't even need a racing car; you raced whatever two-seater transportation you'd driven to the track. If it was unbroken at the end of the race, you drove it home again, too.

This was true even if you campaigned something as potent as a Jaguar D-type or Ferrari Testa Rossa. Since even the fastest sports/racers had to be street legal, they could be and were driven to and from the track. Racers from the Northeast even typically drove their cars to Sebring for the 12 hour endurance race, using the 1200 mile drive as a way to break in their new engine. Of course, nursing a well-thrashed car home from Sebring could be more exciting—and take far longer—than the race itself. Bill Devin, builder of the Devin SS, claims he once bought a Ferrari from Phil Hill in Texas. Hill had raced the car, crashed it, and with no way to get it home, sold it on the spot.

Briggs Cunningham, Lance Reventlow, Tony Parravano, Frank Arciero, John Von Neumann and Vasek Polak were about the only team owners in the whole country with a "rig" to transport their cars, though the Kelso Auto Dynamics van—supported by DuPont money—was considered the wonder of its day when it arrived at Lime Rock in June of 1958. To support the Kelso team of Lister-Chevys, it had been fitted out with a small machine shop, a bedroom and folding stairs that led to the "refreshment deck" located on top of the truck.

By contrast, Elisha Walker, the extremely wealthy owner of the Aston Martin DBR-2 with which George Constantine held the Lime Rock track record, provided mechanic Rex Woodgate with an International Harvester panel truck and open trailer to transport what was probably the most

(Above) Chief starter Terry Field at start-finish. He's standing between the pit lane and the track!
(Left) Ready for the Standing Start, Newt Davis on the pole, Bob Holbert on the outside, Phil Forno in the second row. Yes, you might say F-sports/racing was dominated by Porsche 550s in those days.

(Top) Program from the first Little LeMans race; a Volvo 544 leads a Morris Minor and Renault Dauphine. (Above) The 1958 Saab team with left to right, drivers Gaston Andrey, Dick Thompson, Bob Holbert, Dick Malone, Joe Dodge, Chandler Lawrence, Pup Pupilidy and team manager Dr. John Knight.

LITTLE LE MANS

Of course it was John Fitch's idea; weren't they all? It was simple enough to catch everybody's fancy, too. Lime Rock would host a 12 hour endurance race for small sedans and hopefully all the manufacturers would show up with factory teams. Everybody would have a blast, teetering around and watching the others teeter around and fall off the track and stuff. It would be fun, the manufacturers could use the results in their advertising and everybody would go home a winner. Mostly, it all went according to plan.

There were three classes: 1200 to 1600cc, 750 to 1200cc and 450 to 750cc. Just like LeMans, there was an Index of Performance figured out by Steve Wilder that was calculated using a diabolically complex formula that included the seat to roof measurement of the back seat and the trunk capacity in cubic inches. The only changes from stock were seat belts (no rollbars in 1957), driving lights, tubes in tubeless tires and a choice of sparkplugs. At least two drivers were required per car.

In 1957, 31 cars started the race, which was shortened even before it began by the Connecticut State Police who determined that driving after 6:00 pm in October constituted "night" racing, which was prohibited. So the event became a 10 hour race complete with LeMans start. At the end, five Volvo 444s led four Saab 93s and a lone DKW to form the Top Ten.

Other finishers were Morris Minors, the three-car Renault Dauphine team managed by Rene Dreyfus, a Volkswagen Beetle, etc. At the back were three Panhards. The winning Volvo of Art Riley and Bill Rutan covered 398 laps, two laps ahead of the second place Volvo and 123 laps ahead of the last place Dyna-Panhard.

In 1958, John Fitch moved the race to August so that the finish could be at 8:00 pm in daylight, but now the start was moved back until 10:00 am so it was still a 10 hour race. There were now 34 entries, and amazingly, all but four finished.

After 10 hot and dusty hours, Art Riley and Bill Rutan won just as they had in 1957, with the same private owner Volvo. It was updated with a 4-speed. They finished 399 laps instead of 398, 3 laps ahead of identical Volvos driven by Paul Sagan-Bob Fisher and Charlie Kolb-Dick Wilhelm.

Even though they'd spent a day the week before pre-running the entire event, the best the Saab team could do was fourth overall and first in Wilder's preposterous Index of Performance. They did incite considerable envy and comment by using two-way radios between drivers and pits. Irrepressible Gordon MacKenzie caused even more comment by racing his Fiat 1100 in his trademark kilt. Gordon's kilt was usually just for paddock wear, but on this blistering day, he wore it right into the car.

The 1958 Little LeMans generated an unbelievable amount of press coverage, from *Foreign Cars Illustrated*

(Far left)The Pits during the 1957 Little LeMans race.
(Left) Start of the 1959 Little LeMans. The Team Roosevelt Fiat-Abarth of Roger Penske-Skip Callanan will win.
(Lower left) Volvo P-1300 gets refueled during the 1962 race.
(Left) Bob Holbert runs during a Saab pit stop in 1958.

to *Sports Illustrated*. The whole concept was so nutty, it obviously struck a cord with just about everybody.

In 1959, the same Art Riley-Bill Rutan Volvo was on its way to its third straight victory when, a half-an-hour from the end, two Team Roosevelt Fiat-Abarths caught and passed it for a 1-2 victory. The winners were Skip Callanan and Roger Penske. Abarths won literally thousands of races over the years, but they were certainly never known for their long-term reliability. Little LeMans proved them capable of running flat-out for 10 hours—remember, they were playing catch-up the entire race. It made for the most exciting race of any of the Little LeMans events.

For 1960, the whole concept of the race was changed, mostly because all the American manufacturers had just introduced new compact sedans. Little LeMans was turned into an FIA 8 hour endurance race, which meant, among other things, that all the drivers—most of whom were local SCCA types—had to go through the novelty of obtaining an FIA license. Now they could impress their friends at parties with their "international" racing license, good even at the real LeMans!

The cars were now technically Improved Touring Cars racing under Group II of FIA Appendix J and divided into under-1300cc and over-1300cc categories. There were 31 starters, ranging in size from a pair of Studebaker Larks with 4780cc, 140 hp engines to a pair of Vespa minicars with 400cc and 20 hp. The fast cars were the Larks, Volvos and Saabs.

The Jocko Maggiacomo-Jordan King Lark broke almost immediately after the start, but George Constantine-Ralph Moody kept the other at the front dueling with the Saabs of Gaston Andrey-Dick Thompson and Homer Trotter-Joe Dodge. After 8 hours, the Lark won.

Just about every road racer in the Northeast got a ride in the 1960 Little LeMans, including Bob Akin, Peter Revson, Denise McCluggage, Peter Baumberger, Bob Grossman, Gerry Georgi, Chuck Cunningham, Floyd Stone, Peter Pulver, Frank Baptista and dozens more. Everybody wanted to be on a factory team—even if it was just Vespa, which is where Gordon MacKenzie ended up. Gordon Mac, racing a 20 hp Vespa, whilst wearing his kilt? The mind boggles.

Out of the hundreds of cars and drivers that ran Little LeMans races at Lime Rock, Art Riley-Bill Rutan and their tired old PV-444 Volvo won three races overall—in 1957, '58 and '61—and almost won in 1959. Their secret? The expected combination of excellent preparation, smooth driving and canny race strategy. And of course, a dirt-simple old Volvo that simply couldn't be broken.

Little LeMans entries were off dramatically in 1962. Despite the success of the Riley-Rutan Volvo, it was becoming obvious that only factory-supported cars had a real chance of winning and there were only twenty of those. Volvo was the obvious favorite, racing a 122-S with 1800cc engine called the P-1300. The tired old Riley-Rutan Volvo 444 was also in contention. The Saab and Panhard teams were the other "big guns."

At the end of 8 hours, the team Volvos were first and second, with Art Tattersall-Rune Svensson beating Jack Walsh-Chuck Cunningham. Third was the Saab of Louis Braun-Hal Mayforth, fourth the Panhard of Bob Cory and Howard Humphries.

And that was that. Times changed, local laws changed, the Lime Rock management changed, little sedans changed and maybe racing changed, too. The idea of blasting around Lime Rock in little sedans would have to wait another ten years to be revived in the *Car and Driver* Challenge that started Showroom Stock. After 1962, the Lime Rock Little LeMans endurance race was effectively dead. It's still a great idea, though...anybody for a revival in Showroom Stock and IT cars? It'd be a blast. ■

(Left) Renault team manager Rene Dreyfus.
(Far left) Bob Grossman pushes his hot DKW through Big Bend during practice for the 1957 Little LeMans.
(Lower left) The class-winning Volvo team from 1960: Art Riley, Bill Rutan, Mel Jones and Herb Jeffries.
(Bottom left) In 1958, the Goggomobile of Walter Cronkite-Frank Macauley was the first car out, thanks to a flat battery.
(Below) Dick Thompson in the class-winning Saab, 1957.

(Below) John Fitch and Rene Dreyfus confer in the pits, 1957.
(Right) Shady paddock scene, 1957.

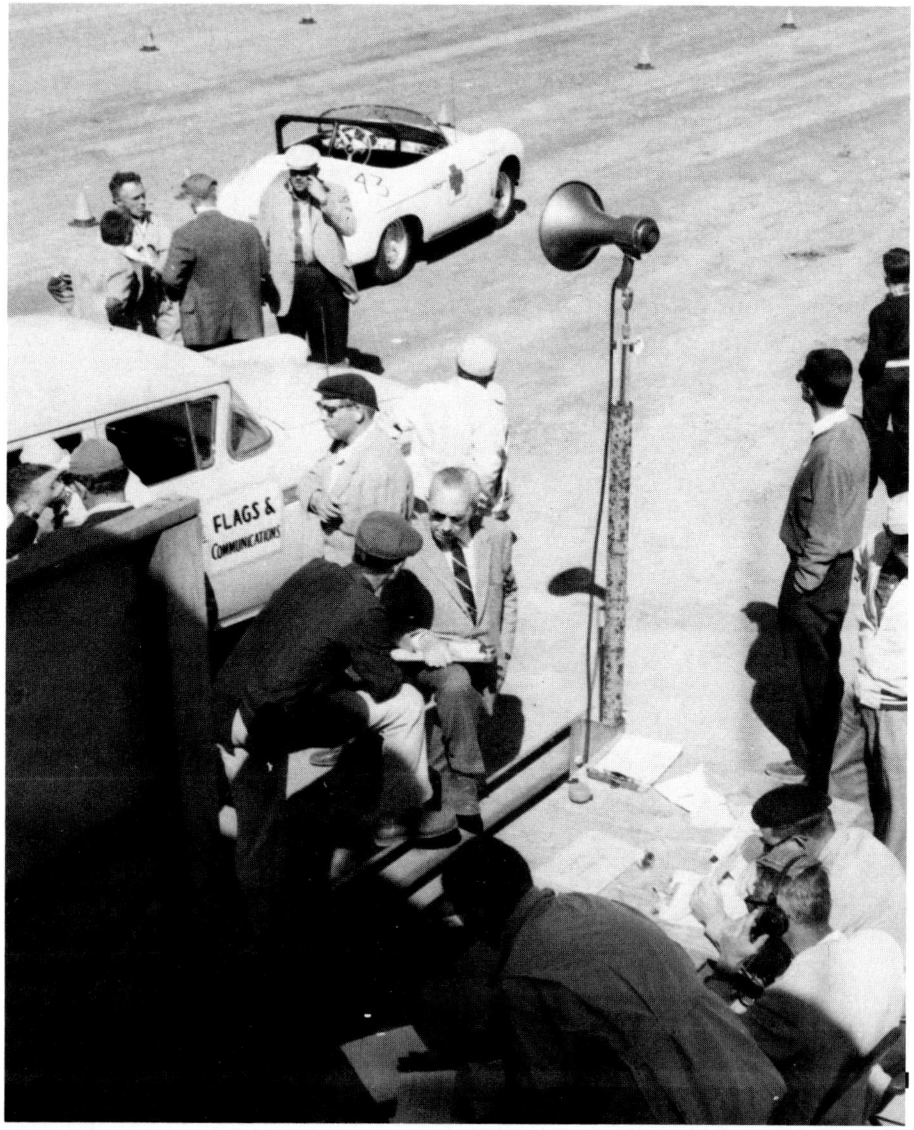

Take another example. When John Fitch raced Vinnie Andrus' 2.0-liter Maserati in 1958, they were among the fastest and best-known teams on the East Coast. Andrus bought the car, John stored it in his barn, a local garage did the maintenance, they towed it on an open trailer behind John's family sedan. Road racing in the Fifties, even at the very highest levels, was less complicated and costly than running a season of SCCA Regionals, today.

You could make a reputation much more quickly, too. Roger Penske, Mark Donohue, Peter Revson—even Grand Prix stars like Jimmy Clark, Graham Hill and Innes Ireland—advanced from their first local event to the highest levels of racing within two or three years. In the late-Fifties and early-Sixties, it was still possible to be literally an overnight sensation.

Less Risky Than War

As Stirling Moss wrote in *My Cars, My Career*, "In the Fifties, many of the people in racing had just survived six years of bloody war, so racing was a piece of cake by comparison. One simply did not consider the risks." Indeed, psychiatrists at the time saw motor racing as a "socially acceptable substitute for combat" that attracted a high percentage of veterans looking for comparable excitement in the peacetime world.

At Lime Rock in particular, there were also a great number of engineers who worked for IBM and who were fascinated by the technical challenges of road racing. Typically, they were far more interested in how the car was behaving than whether you could get hurt while trying to find out.

At the beginning, there were no safety rules at all. Partly as a reaction to Levegh's accident at LeMans in 1955, the SCCA began to tentatively suggest ways to make racing safer. By 1957 when Lime Rock Park opened, roll bars were mandatory for the first time that season, but only in the fastest sports/racing classes. Even that mild requirement had been the subject of acrimonious debate among the SCCA membership.

Production cars—which is what most people raced—were truly "showroom stock." The only safety precautions were removal of the hubcaps, higher air pressure in the tires and a mandatory brake test before practice. Most racers wouldn't have known how to do much more in the way of "race prepping" their race cars.

Most drivers used them, but in 1957, helmets and safety belts were both "suggested," but not required. Drivers *were* required to have "flameproof" driving suits, usually thin cotton poplin soaked overnight in a solution of potassium aluminum sulfate or even sodium bicarbonate, which was supposed to retard burning. In the summer of 1957, Stirling Moss designed the famous light blue overalls which were soon supplied by Dunlop, and which eventually became ubiquitous as racewear around the world.

The SCCA's safety rules were actually far in advance of European regulations. Even Grand Prix drivers were still protected by nothing more than

(Continued on page 41)

(Top) AC-Bristol racers John and Evelyn Mull. (Center) Carroll Shelby on the grid before setting a new track record of 1:06.0 in his 300S Maserati, June 9, 1957. (Bottom) Racing driver/ fashion designer John Weitz. (Left) Ray Saidel testing his latest TVR-based Jomar.

1958

April 26-27: SCCA New England Region, New England Championship. Many familiar names had new cars, including John Weitz with a B-modified Allard LeMans, Gordon MacKenzie with the C-Jag that was destined to find fame as "Mother," and both Denise McCluggage and Newt Davis with Porsche 550 RSs. Gaston Andrey in Mike Garber's 2.0-liter Ferrari Testa Rossa beat John Fitch in Vinnie Andrus' 2.0-liter Maserati in the big race of the day. Denise was third ahead of Gordon. There were 6000 spectators, but only 119 drivers, so the grids were condensed into five 20 lap races. This led to some strange combinations, such as Race Five: B and C-production, Formula III and Formula Libre, with winner Jim Haynes in a Formula III Cooper ahead of Gerry Georgi in the Fitch-Whitmore Special (now Cadillac powered) and Walt Hansgen in a new Jaguar XK-150S.

May 17-18: LISCA. Ran 10 races, each of 15 laps. Among the interesting finishes, Gerry Georgi beat Gordon MacKenzie, Fitch/Whitmore-Cad Special over C-type.

June 14: SCCA New York Region National. Walt Hansgen won his sixth National in a row driving Briggs Cunningham's new Lister-Jaguar, beating Bob Oker's Aston Martin and Bob Holbert's Porsche 550 RS. Jim Jeffords led B and C-production with his "Purple People Eater" Corvette until losing a fan belt. Freddie Windridge in a Corvette won over George Arents in a Ferrari 250 GT.

July 4-5: SCCA National. There were four races of 20 laps each, a 10 lap Formula III race and a 45 lap main event for big-bore modifieds. Harold Hurtley was killed in practice when his Triumph TR-3 overturned on the front straight, his seat belt broke and he was thrown out. He was tied for the National Championship in E-production at the time and ironically, was prominently featured in the race program. Also in the program was an artist's conception of the "John Fitch-Farnum Spicer sports car emporium" to be built in Lakeville. Like many of John Fitch's good ideas, it never happened. Ed Crawford won the big-bore race over Walt Hansgen, both in Lister-Jaguars.

July 12-13: LISCA.

July 26-27: Westchester Sports Car Club.

August 2: Second Little LeMans Endurance Race. Some 34 under-1600 sedans started the 10 hour grind, and 30 finished. But just as in 1957, Bill Rutan and Art Riley won with the same Volvo, covering one more lap than they had the year before. Volvos were also second and third, with the Bob Holbert-Gaston Andrey Saab fourth overall and first in Index of Performance.

September 6-7: USAC First Annual Lime Rock International Open. George Constantine, driving Elisha Walker's Aston Martin, both won this first USAC professional road race in America and set a new Lime Rock lap record at 1:05.81.

October 18-19: New York Region Championship.

A wonderful shot of the paddock during the first USAC International Open, September 7, 1958. The steep dirt road down from the hill was also the access road to Jim Vaill's gravel pit, just out of sight to the right. There are still an impressive number of trees everywhere, and even thick brush in the middle of the parking area. In addition to the concession stand and some exotic street machines—including two Ferrari 250 GTs—Elisha Walker's Aston-Martin DBR-2 #49 is in plain sight next to its tow rig. Remember, this was the top-line team at the time, and driver George Constantine broke the lap record at 1:05.8 later this same afternoon. They've got an International Harvester panel truck, an open trailer with a partial cover and a tarp tied to tent pegs. The average regional racer has a fancier rig, today. I mean, like, where's the motorhome, Dude?

1959

April 26: SCCA New England Region. Postponed because of rain.

May 2: LISCA.

May 9: SCCA New England Region rain date. New ban on Sunday racing; Saturday morning practice, six 20 lap races starting at 2:00 pm Saturday. Among the other big changes from 1958, Stutz Plaisted had parked his homebuilt "Cheetah" in favor of a new MGA, Bob Grossman, Floyd Stone and Charlie Rainville were all racing Alfa Veloces, Bill Wonder had an AC-Bristol, Bob Bucher a Lotus Eleven Clubman, Denise McCluggage an OSCA, Candy Poole his famous PBX and J.D. Igleheart a Nardi. John "Bunky" Hearst, in his Deutsch-Bonnet coupe, won the first race he ever entered. Gaston Andrey won the "big-bore" contest with his 2.0-liter Ferrari Testa Rossa ahead of Farnum Spicer's 1.5-liter Maserati. Other race winners were Plaisted, Rainville, Harry Blanchard (Porsche Carrera), Gerry Georgi (Cad-Fitch/Whitmore Special) and Denise.

June 6: USAC International Formula Libre. Postponed for lack of entries.

June 13: SCCA New York Region. Postponed because of rain.

July 4: SCCA National. Among the unfamiliar entries were John "Skip" Barber in an Austin Healey Sprite, Richard Bauer in an MG-TD, Frank Dominiani in a Corvette, Timmy Mayer in an Austin Healey and Roger Penske in Bob Holbert's old 550 RS Porsche. George Constantine and Walt Hansgen provided the big excitement until Hansgen's Lister broke, leaving Constantine's Aston DBR-2 ahead of Briggs Cunningham in the other Cunningham team Lister-Jaguar.

July 11: SCCA New York Region rain date. Roger Penske in the Porsche won the big race overall, ahead of Floyd Aaskov's Lotus Eleven.

July 25: USAC International Formula Libre. Probably the most famous race—and maybe the best—ever held at Lime Rock Park. 1959 Indy 500 winner Rodger Ward, driving an old Offy-powered 1.7-liter Kurtis midget, set a new lap record of 1:04.67 and handily beat George Constantine's Aston Martin DBR-2, Chuck Daigh's ex-Fangio Maserati 250F Formula One car, Pedro Rodriquez' Maserati 300S and John Fitch's Cooper Monaco. Constantine was credited with the absolute track record at 1:04.83, because Ward's time was set in practice, not during a race.

August 15: Third annual Little LeMans. Team Roosevelt Fiat-Abarths finished 1-2, passing the Art Riley-Bill Rutan Volvo just half-an-hour from the end. The winners were Roger Penske-Skip Callanan.

September 12: LISCA. Among the neophyte drivers were Charlie Gibson (Sprite) and Peter Pulver (TR-3).

October 17: SCCA New York Region. More neophyte drivers, including Bob Akin, III in an Alfa Veloce and Sherm Decker in an MGA Twin-Cam.

October 31: Westchester Sports Car Club. Their motto "A class for every car and a prize for every class."

a shortsleeved Suixtil or LaCoste tennis shirt as late as 1961. And on both sides of the Atlantic, goggles and gloves were worn not for any reasons of safety, but because they gave tear-free vision at high speeds and a better grip on the polished mahogany steering wheel, respectively.

The Cars They Drove

Between 1948 and 1955, there was really little improvement in the sports cars that gentlemen raced. In the early days, there were only two SCCA classes; Under 1500cc and Over 1500cc. There was no division into Production and Sports/racer as there was later on, and the Sports Car Club of America raced two-seater cars with fenders. Open-wheel single-seaters were not welcome in those days because they were not sports cars.

Nobody knew much about how to improve the performance of their cars, so just about any modification was allowed. The architypical sports car in the early days of the SCCA was an Allard. An Allard is essentially a '32 Ford hot-rod with a Cadillac V-8 and a funny-looking aluminum body.

(Continued on page 51)

(Above) September 20, 1958: unprotected teenage corner workers next to the track, somewhere around the outside apex from West Bend. And sitting down, besides!
(Far left) Corner worker Bill Chatfield gets an errant Austin-Healey back into Big Bend in 1959.
(Upper left) Racing enthusiasts Dave Garroway and Skitch Henderson in 1959, with Garroway's famous Jaguar SS-100.
(Bottom) Drivers' Meetings never change. This is 1962.

FORMULA LIBRE

On July 25, 1959, Lime Rock Park hosted what John Fitch calls "the best race ever held by anybody, anywhere." Not only the *best*, but the most memorable. Not only the most memorable, but the most off-beat. Sanctioned by USAC and sponsored by the Hearst tabloid *New York Mirror*, the concept was revolutionary. Run watcha brung. For money.

Actually, this was the third USAC road race in the east...there had been one at Lime Rock in September of 1958, another at Watkins Glen. Both were aesthetic and financial disasters. The Formula Libre race was headed the same way. Originally scheduled in June, it was postponed a month by USAC competition director Duan Carter and Lime Rock director John Fitch.

The extra month gave them time to put the arm on everyone they knew. Rodger Ward—fresh from winning the Indy 500—was expecting to drive a new Cooper Monaco owned by Rambler dealer Charles Kreisler. When Fitch got the ride instead, announcer Chris Economaki suggested to Ward that a tiny Offy-powered midget might be the secret weapon around Lime Rock. Ward agreed to drive one owned by Kenny Brenn from New Jersey. Though immaculately prepared, Brenn's Kurtis was already 11 years old and had raced over a thousand times. Reigning USAC National Champion Tony Bettenhausen, Duane Carter and local driver Bert Brooks also showed up with similar midgets.

These dirt track Kurtis midgets had four-cylinder, 1.7-liter Offenhauser engines, a 1-speed "In and Out" gearbox, treadless slicks and rear-wheel-only brakes operated by an outside hand lever. Nothing like them had been seen on a road course since before the Kaiser War.

The competition was a mixed bag, to say the least. Scarab star Chuck Daigh drove Lucky Casner's Camoradi Maserati 250F, Juan Manuel Fangio's 1958 Formula I car, now painted white and blue. Daigh's normal patron, Lance Reventlow, brought his own Cooper Formula II. The only other open-wheeler was Ray Saidel's homemade Jomar Formula II.

There were also a dozen sports/racers of various types and sizes. The best-known at the time was Lime Rock track record holder George Constantine with Elisha Walker's Aston Martin DBR-2. John Fitch had Kreisler's Cooper Monaco, teenager Pedro Rodriguez had a Maserati 300S, Gordon MacKenzie his C-type Jaguar and Dick Thompson had Bill Mitchell's famous Stingray. *Competition Press* editor Denise McCluggage borrowed a Porsche 550 RS from Briggs Cunningham.

A record-breaking crowd of nearly 12,000 showed up to watch their sports car heros blow the Indy 500 winner into the Lime Rock weeds. Imagine their surprise when Ward in the tall, old-fashioned midget set a new unofficial track record in qualifying at 1:04.67 and started from the pole for the first 20 lap heat. George Constantine out-accelerated him at the start, however, and won with Ward, Daigh, Fitch and Rodriguez behind him in that order.

Before the next 20 lap heat, Ward switched to larger 16 inch rear wheels, for more top speed on the straight.

(Background) George Constantine's Aston Martin leads Chuck Daigh's Maserati and Rodger Ward's midget through Echo Valley.
(Top left) Dick Thompson in Bill Mitchell's Stingray.
(Center) George Constantine in Elisha Walker's Aston Martin DBR-2.
(Bottom) Lance Reventlow in his Formula II Cooper-Climax.

He and Constantine diced throughout, with Ward finally winning the second heat by 4 seconds over Constantine, Daigh, Fitch and Rodriguez.

In the third heat, a 60 lapper, Constantine led Ward until Chuck Daigh passed both of them with the Maserati. Rodger Ward then brought the crowd to its feet by pulling an "impossible" move and passing both Daigh and Constantine in one sweep around the outside The Hook. Soon after, the Aston spun a rear wheel bearing and retired.

Ward had one more spectacular way to get the crowd on its feet again. Coming out of The Hook, just in front of Daigh's Maserati, Ward got wide and into the rough. Instead of spinning the way a road racer would do, the old dirt tracker kept his foot in it and broadsided the midget in a delicious four-wheel drift across the dirt, intersecting the track a couple of hundred yards later when it turned left in the midst of The Esses. He was *still* ahead of the amazed Daigh. The crowd went wild.

Daigh led for a while, but Ward passed him with 15 laps to go and won the third heat from Daigh, Rodriguez, Fitch and Colombosian. When all three heats were totalled

(Right) Chuck Daigh in the ex-Fangio Camoradi Maserati side-by-side with Rodger Ward in Kenny Brenn's Kurtis Kraft midget coming through The Hook. Imagine the crowd's surprise when the little midget blew by Daigh's Formula I car.

up, Ward had obviously won overall.

There has never been a race like this, before or since...cars of all sorts, with and without fenders, from 1.5 to 4.9-liters, air-cooled and water-cooled, front and mid-engined, slicks and treaded tires, even running on different fuels. The midgets used alcohol, the Maserati Avgas, the sports cars gasoline.

Neither USAC nor Lime Rock successfully staged a similar race again...Formula Libre was a once-in-a-lifetime event. Today, when half the IndyCar races are on road courses and most of the drivers got their start with the SCCA, it's hard to remember that it wasn't always this way. In mixing road racers and oval track stars, Lime Rock's famous Formula Libre was decades ahead of its time.

A footnote about "horses for courses." Rodger Ward was so encouraged by his showing at Lime Rock against world-class competition that when the first United States Grand Prix was held at Sebring in the autumn of 1959, he entered his own USAC Offenhauser-Kurtis. What worked at Lime Rock's bullring was all wrong for the endless airport runways of Sebring's 5.2-mile course. Ward got his doors blown off. ■

1959 USAC International Formula Libre (Final Results)

1. Rodger Ward, 1.7-liter Offenhauser-Kurtis
2. Chuck Daigh, 2.5-liter Maserati 250F
3. Pedro Rodriguez, 3.0-liter Maserati 300S
4. John Fitch, 2.0-liter Cooper Monaco
5. Bob Colombosian, 2.0-liter Lister-Bristol
6. George Constantine, 4.2-liter Aston Martin DBR-2
7. Lance Reventlow, 1.5-liter Cooper Formula II
8. Bert Brooks, 1.7-liter Offenhauser-Kurtis
9. Dick Thompson, 4.9-liter Stingray
10. Russ Klar/Tony Bettenhausen, 1.7-liter Offenhauser-Kurtis
11. Jocko Maggiacomo, 4.9-liter Chevrolet-Ferrari
12. Denise McCluggage, 1.5-liter Porsche 550 RS
13. Ray Saidel, 1.5-liter Jomar Formula II
14. Loyal Katskee, 3.0-liter Ferrari Monza
15. Gordon MacKenzie, 3.4-liter Jaguar C-type
16. John Grimaldi, 3.4-liter Jaguar D-type
17. Duane Carter, 1.7-liter Offenhauser-Kurtis
18. Paul O'Shea, 4.6-liter Bocar XP-5
19. Vic Meinhardt, 1.5-liter Porsche 550 RS

(Left) Movie star-handsome Lance Reventlow—son of Woolworth heiress Barbara Hutton, "the richest woman in the world"—at Lime Rock during the Formula Libre weekend. His fast and lovely Scarabs unfortunately never raced at Lime Rock, but they sure brought a lot of attention to American road racing specials in the late-Fifties and continued winning into the mid-Sixties.

GORDON MACKENZIE

He's the very embodiment of the gentleman amateur, a wonderful character who certainly put as much back into the sport as he took out. Gordon MacKenzie was the son of the local doctor in fancy Millbrook, NY, but he never really had much money. He spent his career as a manager for IBM in Poughkeepsie, and owned only two race cars during his whole career.

The first was a Jaguar XK-120M that he ran from 1953 until 1957. In the middle of the 1957 season, Gordon showed up with an old C-type Jaguar, which he promptly christened "Mother," and with which he ran C-modified for the next eleven years. Both of Gordon's cars were out-of-date before he started racing them, cars that other drivers had already abandoned as uncompetitive. His lifetime record? Gordon MacKenzie appeared in 176 races, won 30, finished second 45 times and third 29 times. He had only 16 DNFs in all those years. And remember, by the end he was racing a C-type against mid-engined McLarens and stuff like that. Truly a remarkable racing driver.

But an even more remarkable man. Gordon was New York Region executive, he was a perennial teacher at SCCA driver's schools, and with his MacKenzie kilts, one of the "sights" of Lime Rock that inevitably ended up in every newspaper race report.

Gordon ran his one-man equippe with admirable professionalism, and a lot of the information in this book came from his meticulously organized files covering every event he attended. Now retired from IBM, Gordon splits his time between an island in Maine, a sailboat in the Caribbean and instructing at VSCCA drivers schools. Forty years after he started, Gordon still has his foot to the floor. ■

(Opposite page) The paddock during the Formula Libre race.
(Below) Kilted MacKenzie at the original Lime Rock gas pumps on Opening Day, 1957.
(Lower left) "Mother" leading Gerry Georgi in the Cadillac-powered Fitch/Whitmore Special, May 18, 1958.

(Overleaf) The very definition of a gentleman amateur racing driver, Gordon MacKenzie in his famous Jaguar C-type, known affectionately as "Mother," was a front-runner at Lime Rock for a decade in a car that was already obsolete when he started.

52

position of leading the revolution trying to break the power of the SCCA, yet at the same time holding SCCA amateur races every weekend. As he admits rather ruefully, "There were lots of battles through those years."

Race Fever

(Above) Harvey Siegelman on the Lime Rock pre-grid in his Bandini Formula Junior in 1961.

Lime Rock was also part of another revolution in road racing. The first race at Lime Rock was April 28, 1957. The first race at the new 3.0 mile circuit at Bridgehampton, on Long Island, was on May 12. The first race at the new 2.0 mile circuit in Thompson, Connecticut was on May 26. The first race at the new 2.0 mile circuit in Marlboro, Maryland was on July 14.

It was a true revolution. In a matter of months, East Coast road racers had gone from racing at airports like Montgomery, New York or informal hillclimbs like the Cloudbank Hill Climb up a private driveway in Garrison, New York to having their pick among these new road racing circuits, all designed expressly for the purpose. These new circuits, particularly Lime Rock Park, were infinitely safer and more interesting upon which to race than the tree-lined street circuits or rough airports that had been the only choices before.

Road racers didn't get around that much in the Fifties. For one thing, there were few Interstate Highways, so even a trip to the Midwest could be a grueling three-day tow on two-lane roads. Going all the way to California was unthinkable. European racing was something you knew about from reading Bernard Cahier in *Road & Track*. West Coast racing, the same.

East Coast racing was covered almost exclusively by Manhattan-based *Competition Press* and local newspapers. When Carroll Shelby and Bruce Kessler showed up at Lime Rock in 1957 with race cars that had come all the way from California, it was a major event, mostly because we'd read their names in the magazines. That put them in the same category with Grand Prix stars; racing drivers who got their names in print. Wow!

On Any Saturday

Before all the new tracks opened, East Coast racers might only get to compete half-a-dozen times a summer. Now there was a race every weekend—heck, there was a race every weekend just at Lime Rock—that all could be reached from anywhere in the Northeast with only a few hours drive from home.

This had two important results. One, there was a tremendous boom in road racing enthusiam among the racers, spectators and press alike. In the late-Fifties, it was nothing for 10,000 spectators to show up at Lime Rock for an SCCA National. Two, the quality of cars and driving rapidly improved and the size of the grids increased enormously. Instead of running three or four races mixing all sorts of classes together with perhaps only twenty cars in each race, now Lime Rock was holding eight or nine races on a Saturday with thirty or thirty-five cars in each race. It truly was a Golden Era in East Coast road racing.

"Lime Rock Experts"

Every race track has its own personality. Some are sensible and easy to learn, Willow Springs, for example. Some are confusing like Sebring. Some are awkward and graceless, like Bridgehampton. Some, like Watkins Glen, have a beautiful rhythm that is fast, smooth and fun. And then there's Lime Rock.

To drive around Lime Rock is easy. To race around Lime Rock is a little more difficult. To go truly fast around Lime Rock is extremely difficult, indeed. There is no more subtle race

(Continued on page 65)

FORMULA JUNIOR

Count Giovanni "Johnny" Lurani, that most cultured of Italian racing driver/journalists, came up with the idea in 1957. He proposed a "Junior Formula" based on Fiat 1100 components to replace Formula III as a starter series for young racing drivers. Roberto Lippi won the first Italian Junior Championship in 1958. By the 1959 season, FJ was an FIA class and people were enthusiastically building and racing them worldwide. The old motorcycle-powered Formula IIIs were almost instantly eclipsed.

The FJ rules were for single-seater racing cars using production car components. Cars with engines up to 1100cc had to weigh at least 880 lbs. Cars under 1000cc got a weight break to 792 lbs. Overhead cams were prohibited, and running gear had to be from the same car as the engine. Very innovative at the time, a rollbar was required.

Most FJs looked like miniature Formula I cars; the Fiat-based Stanguellini, Lancia and OSCA were tiny Ferrari clones; the BMC-based Lola, Elva and Gemini looked like Vanwalls.

The revolution in Formula Junior was the multi-purpose 1959 Lotus 18, the first mid-engine Lotus. It could be fitted with everything from a 948cc BMC for FJ up to a 2.5-liter Coventry-Climax for F/I. This was followed in 1961 by the Lotus 20/22, and in 1963 by the monocoque Lotus 27.

That was the end of Formula Junior; by 1963, it had become a factory-supported class for testing Formula I drivers and technology rather than a starter class for amateurs. For 1964, FJ was replaced with Formula II and Formula III. In the States, old FJs were fitted with a variety of engines to become Formula B and C cars. Many raced right up until the Seventies. ■

(Top left) Taraschi was a typical Italian front-engine Formula Junior. This is Bob Girvin in the Lime Rock BMW Vintage Fall Festival of 1985, though except for his helmet, it could be 1959. (Bottom left) The mid-engine 1961 Lotus 20 revolutioned Formula Junior. This is Harry Carter at Lime Rock on his way to winning the SCCA FJ National Championship in 1961.

1962

June 30: SCCA New England Region National. 206 entries in 8 races starting at 11:00 am (practice and qualifying on Friday). Among the emerging stars; Randy Canfield in a Sprite, Skip Barber in a Turner, Art Tattersall in an MGA, Gerry Sagerman in an Elva, Bob Tullius in a TR-4, R. Harry Gaunt in an Elva, Michael Gammino and Dick Lang in Corvettes, William O'Donnell in a Lotus Super 7 and Eno DePasquale in a Lister-Buick. The big star of the day was Roger Penske. Driving his red Cooper Monaco, he set a new track record at 1:03.4, led every lap of the main event and lapped the entire field, including Bob Holbert's RSK in second. Timmy Mayer set a class record on his way to winning Formula Junior, then was disqualified for his Cooper being underweight. In the production race, Gerry Georgi in Allan Wylie's Ferrari Berlinetta beat Bob Grossman's identical car by 4 seconds.

July 7: Sixth Annual FIA International Little LeMans. Only 20 cars started the 8 hour grind this year, and the first car to quit, an Austin Mini-Cooper driven by Denise McCluggage, was already on the trailer before Denise's co-driver Bob Grossman had time to get into his driving suit. At the end, two factory Volvo 122-Ss won, driven by Rune Svensson-Art Tattersall and Chuck Cunningham-Jack Walsh. Louis Braun-Hal Mayforth were third in a Saab.

July 21: LISCA. LISCA classes were different from the SCCA and graphically show the changing state of racing in the early-Sixties:

1. NSU Sport Prinz, Fiat 600, Triumph Herald, MGTD, etc.
2. Austin Healey Sprite, Fiat-Abarth, Turner Sports 948, MGTF 1500, etc.
3. Porsche 1300, Alfa Romeo Giulietta, Morgan 4/4, Volvo, Saab GT, etc.
4. Sunbeam Alpine, Volvo P-1800, Mercedes 190SL, MGA, Facel Facellia, etc.
5. Morgan +4, Elva Courier, Austin Healey 100, Triumph, Porsche 1600, etc.
6. Austin Healey 3000, AC Ace, Lotus 7-Ford, etc.
7. Mercedes 300SL, Aston Martin DB2-DB4, Jaguar XK-120-XK-150, Siata 208, Daimler SP-250, etc.
8. Jaguar XKE, Ferrari 250 GT, Corvette, Porsche Carrera, etc.
9. Modified cars under 2000cc, front-engine Formula Junior, Formula III.
10. Modified cars over 2000cc, plus Lotus 15 and 19, Porsche RS and RSK, rear-engine Formula Junior.

August 25: SCCA Northern New Jersey Region. New names include Freddie Ashplant, David Rivkin, Alex Dearborn and Herb Wetanson.

October 20: SCCA New York Region.

GASTON ANDREY

"I first came to America as an engineering student in 1949," says Gaston Andrey. "I spent a year at Studebaker in South Bend working on the first V-8. Then I went back to Switzerland, did my military service, and came back to America in 1952. And I started a little foreign car business."

That little foreign car business grew to four dealership locations. "Then I sold out the lot, and kept just one location," he says. Today, Gaston Andrey West in Framingham, Massachusetts sells Ferrari, Alfa-Romeo and Saab. Typical of Gus Andrey, these are all car companies with which he has been happily associated for more than three decades.

He started racing in 1953. "It was very informal," says Mr. Andrey. "I drove a Jaguar special that had been put together by Walt Hansgen and was driven by him and Paul Timkin. Then I got some customers who were racing, and they let me drive their cars at Thompson. I didn't do too badly, so they let me race their cars. They got the trophy and I got to drive! That was fine with me."

The big turning point for racing driver Gaston Andrey came when he met Mike Garber, owner of a prosperous New England chemical company. Garber supplied him with a Siata V-8 followed by a Ferrari Mondial, a 2.0-liter Testa Rossa and then a Maserati Tipo 61. After that, there was a Formula Junior Lotus 18, followed by a Lotus 20/22 that was eventually converted to a Formula B car with an Alfa engine. In the late-Sixties/early-Seventies, Gaston partnered with Horst Kwech, driving Alfa-Romeo GTAs in the under-2.5 Trans-Am.

All together, Gaston Andrey won National Championships seven times—once in the Mondial, twice in the Testa Rossa, twice in the Maserati and twice in the Trans-Am Alfa. He kept racing at the top level for over three decades; his last race was an IMSA race at Miami in 1985 when he crashed heavily into the wall and as he says, "Totally demolished the car and myself. After that I decided 'That's Enough.' The bones only heal so much. After that I only drive in some vintage and historic cars, my old Maserati and cars like that." ■

(Left) Gaston Andrey in 1958.
(Center) At speed on the Lime Rock front straight in Mike Garber's 2.0-liter Ferrari, 1958.
(Bottom) Gus in an obsolete Lotus 18 about to pass a new Lotus 20. Those who remember the difference in performance between these two cars will be very impressed, indeed. Andrey went on to win this Formula Junior race at Lime Rock in 1961.

BRITISH MOTOR TRIALS

Since 1948, the British Motoring Guild has held an annual "press day" at the Goodwood race track in Sussex. Writers and broadcasters get a chance to drive and photograph all the new cars and report back to their readers. This tradition goes on in this country, too, with the annual Test Day of the International Motor Press Association. This was held for years at Lime Rock in October and only recently moved to Pocono International Raceway.

It's hard to remember, but back in the Fifties—before IMPA even existed—British manufacturers built and sold the majority of the cars imported to this country. Particularly performance cars. So in 1956, the British Automobile Manufacturer's Association organized the British Motor Trials, a sort of Americanized version of Goodwood Press Day.

On May 25, 1957, the second annual British Motor Trials was held at newly-opened Lime Rock Park. There were thirty-five different cars to try, ranging from a Morris Minor to a Jaguar SS, the famous street-legal D-type. Over 200 reporters, plus 300 hangers-on, drove around all day, picnicked on the grass and finished up with a cocktail party in the paddock. There was even a celebrity time trial with Skitch Henderson, Paul Whiteman, Bert Parks, Walter Cronkite and Jack Lescoulie plus demonstration runs by the three MGs that had won the team prize at Sebring in March.

Similar British Motor Trials were held each year into the mid-Sixties, when the event evolved into IMPA Flaunt Your Incompetency Day with cars from all manufacturers. The highpoint of all the press days at Lime Rock came in 1961, when BMC invited not only reporters and photographers, but a group of Grand Prix drivers who took part in a celebrity race of identical Austin Minis.

This would have been just another forgotten publicity stunt except that with Lime Rock regulars Denise McCluggage and John Fitch, and Pedro Rodriguez who'd raced here before, BMC brought Juan Manuel Fangio, Innes Ireland and Masten Gregory to Lime Rock. This was the only time Fangio ever visited the track, the only time Gregory ever visited the track and the only time for another two decades that Innes Ireland would be here. The stars went out and played around in their identical little cars—sort of an early-day IROC—posed for the photographers wearing funny BMC berets and then repaired to the bar.

It would have been just another corporate entertainment day—Lime Rock hosts dozens of similar events each year—except for the presence of Fangio and his friends. And that makes it a very special day to remember, the day Juan Manuel Fangio raced at Lime Rock...albeit in a Mini. ■

(Opposite page) Innes Ireland buckles up in anticipation of one of his more exciting afternoons of motor sport.
(Top left) Leaning into Big Bend, the Mini grid promises a dicey race to come.
(Far left) Juan Manuel Fangio and Masten Gregory. This was Fangio's only appearance at Lime Rock.
(Near left) Innes Ireland and Pedro Rodriguez. If this was a magazine from 1961, the caption would continue "discuss strategy." Asked what they were really talking about in all those shots, Stirling Moss replied "Crumpet, of course."

1963

May 4: LISCA. Jim Haynes, driving his Lotus 18 Formula Junior, set a new Lime Rock track record at 1:02.9.

June 15: SCCA New England Region National. The new SCCA Production Categories included:

Class A: AC Cobra, Aston Martin DB-4 GT/Zagato, Corvette Sting Ray, Jaguar XKE, etc.
Class B: Ferrari 250GT, Lotus Super 7, Lotus Elan, Porsche Carrera, Mercedes 300SL, etc.
Class C: AC Bristol, Alfa Romeo Sprint Zagato, BMW 507, Morgan +4 Super Sports, etc.
Class D: Arnolt Bristol, Austin Healey 3000, MGB 1800, Triumph TR-4, etc.
Class E: Elva Courier 1600, MGA, Porsche 1600, Sunbeam Harrington LeMans, etc.
Class F: Lotus 7-A, Mercedes 190SL, MGA, Denzel 1300, Fairthorpe Electron, etc.
Class G: Fiat 1500, Morgan 4-4, Triumph Spitfire, etc.
Class H: Austin Healey Sprite, Fiat-Abarth 750 GT, MG T-series, etc.

September 28: SCCA Northern New Jersey Region. Among the new names: Fred Opert in a Sprite, Harry Theodoracopulos in an Alfa Ti sedan, Dick Gilmartin in a TR-3, George Wintersteen in a Porsche RSK, Skip Scott in an Elva.

October 19: SCCA New York Region. Another newcomer; Jerry Walsh in a Sprite.

George Alderman set a new Lime Rock track record at 1:02.0 with his Cooper Formula Junior.

track in the world, which of course, is what Jim Vaill and John Fitch intended when it was laid out. Thousands of racers have proven over the past four decades that the only way to get really fast at Lime Rock is to put in a lot of laps.

Which brings us to the Lime Rock Expert. In every period, there have been a handful of drivers who've mastered the subtleties of Lime Rock and been able to get around it more quickly than their contemporaries in identical cars. Some of these drivers went on to become international figures who went fast at any race track. Others were competitive only at Lime Rock. Still others were competitive everywhere, but comparatively more competitive at Lime Rock than other places. Among the acknowledged Lime Rock experts in the late-Fifties and early-Sixties were locally-famous names like Harry Carter, Newt Davis, Jim

Bob Johnson from Columbus, Ohio in his 289 Cobra, at Lime Rock on June 15, 1963. He finished third in the SCCA National. This car has many distinctive touches, including home made "spats" behind the front wheelwells.

Haynes, J.D. Igleheart and George Constantine, as well as better-known stars like John Fitch, Gaston Andrey and Roger Penske.

The Passing of the Gentlemen Amateurs

The ideal of the gentleman racer died around 1963. Why? Cynics will say there were no more gentlemen left, but it's more than that. As racing became more popular, it also became more competitive. There were now people willing to spend an inordinate amount of time and money to win a pot metal trophy. At the same time, both production cars and sports/racers became not only more specialized, but both infinitely adjustable and expensive.

By 1963, a Cooper or Lotus was up in the $20,000 range, a Maserati or Ferrari was much more. That was a lot of money back then. Worse, these cars required a knowledgeable mechanic to maintain them in fighting trim. The concept of "setting up" a race car chassis dates only from the early-Sixties. Before that, you checked the oil, the gas and the tire pressures. If the car didn't work properly, you drove around its deficiencies. If that wasn't enough, you bought another car. The idea of painstakingly sorting out the handling just wasn't part of what gentleman racers did.

John Fitch opened the first racing driver's school in America at Lime Rock Park in 1957. He got together a couple of economy sedans and showed neophyte racers the proper way to shift, brake and turn corners. Such basic concepts as trail braking, the friction circle and a square chassis were totally unheard of, even by world-class racers. This was not an isolated phenomenon. Fon Portago drove at the international level for more than a year before Maserati engineer Guerrino Bertocchi taught him the proper way to shift.

By 1963, such ignorance was inexcusable, even in a Spanish grandee. There had grown up a cadre of professional and semi-professional drivers and mechanics who not only understood something about vehicle dynamics—including aerodynamics—but who were starting to convince large corporations that business could be done by painting product names on the sides of race cars in return for a monetary consideration.

In sum, the sport was changing into a business, and gentlemen amateurs were being left behind. The cars had changed, the people had changed, the whole atmosphere had changed. All that remained the same was Lime Rock Park. And even that was involved in a bitter civil war. Racing is not always easy.

3.
Going Pro

HAGGAR slacks
HAGGAR
6
BRETT LUNGER
BILL MAYBERRY
HOGAN RACING LTD.
GOODYEAR
slacks
HAGGAR

(Previous page) Sponsored by Haggar slacks and owned by Carl Hogan—Bobby Rahal's IndyCar partner today—in 1972 Hogan Racing headquartered across Route 112 from Lime Rock Park in what is now The Paradise Garage. Bill "Murph" Mayberry was in charge of the team's Formula 5000 Lola T-300-Chevys, driven by Brett Lunger and David Hobbs.

The Sixties was a period of great turmoil for America in general, for American racing, and for Lime Rock in particular. Just as in the country at large, everything to do with racing, from the design of the tires to the running of the race tracks, underwent great change.

Racing became significantly more professional during these years, exchanging its old amateur ideals for unfamiliar business demands. This was a fundamental societal change. You can make a pretty good case that when John Fitch and Briggs Cunningham retired after racing at Sebring in 1966, the "gentleman amateur" racing driver—after years on the endangered species list—was officially extinct.

Turmoil at the Track

It's no secret that Lime Rock Park was under attack even before the first race was run in 1957. A small but vocal group of local residents tried from the start to close the track, even though the town of Lime Rock had voted overwhelmingly in favor of building it.

When that didn't work, the dissidents invested two years in getting a court injunction that put racing in the Never On Sunday category, restricted the hours of operation from 9 am to 6 pm for unmuffled cars and allowed only ten "unmuffled" weekends of racing per year.

Ironically, a lot of people think that Saturday Only racing is one of the things that has made Lime Rock so popular over the decades. There is no more convenient, enjoyable place to go racing in the United States. And as thousands of racing drivers have pointed out to their skeptical wives, on Sunday you get to do something else—"Something that *you* want to do, Dear!"—that has nothing to do with racing.

It's also no secret that until comparatively recently, Lime Rock Park was always underfinanced. What few people know, however, is that track builder Jim Vaill never owned the farm/gravel pit on which the track was built. The Lime Rock Corporation owned the track itself, but the property was leased to the corporation for five years by Jim's father, Frank Vaill. At the beginning, this was a great advantage to Lime Rock Park, because Jim could spend his limited resources building the track, not acquiring land. But in 1962, it backfired on him.

Simultaneously, the original lease ran out and Frank Vaill found himself short of cash for another project. He offered the 385 acres of Lime Rock Park to The Lime Rock Corporation for the bargain price of $50,000. The corporation didn't have ten dollars, let alone fifty grand. Mr. Vaill subsequently sold the property to local resident John Norwood, who was a creditor of the corporation.

Sadly, this precipitated a fight among The Lime Rock Corporation stockholders that was so bitter nobody really wants to talk much about it, even today. Without getting into personalities, there was a minority stockholder suit against the corporate management and in an out-of-court settlement, the minority shareholders took over the management of the track.

New Owner

The May 2, 1964 edition of the Lime Rock Park track program opens with an article entitled "Jim Haynes Obtains Control Of The Lime Rock Corporation." The Proposed Directors of the reorganized corporation were Jim Haynes, John Fitch, Newton Davis and attorney William Wilbourne, III. As Jim Haynes says, "I never owned the track as an individual, though I did own or control the majority of the stock."

Very soon, John Fitch and Newt Davis moved away from active involvement with the track, leaving Jim Haynes as sole track manager, President and Treasurer of The Lime

(Continued on page 71)

JIM HAYNES

Jim Haynes' grandfather was Jacob Haynes, as in the Haynes and Haynes-Apperson cars built in Kokomo, Indiana between 1898 and 1925. Originally from Port Huron, Michigan, Jim came east to Yale. He graduated in 1955 with a degree in Industrial Engineering and went to work for IBM in Poughkeepsie. In 1960, he left IBM and founded Lakeville Precision Moulding, maker of tiny plastic gears and machine components.

Haynes' first sports car was an MG-TD, which gave way in 1957 to a Morgan +4. After a season of racing, the Morgan became the tow car for Jim's Cooper-Norton Formula III. Jim Haynes was 1959 SCCA National Champion with a more advanced Cooper-Norton Mark XIII. During the winters, he raced Three-Quarter midgets for money, indoors, under the *nom de course* of "Eddie James" so the SCCA wouldn't throw him out for being a professional driver.

After the Cooper, he had a Formula II Cooper-Climax, then he and Jocko Maggiacomo built the ill-fated Jocko Formula Junior. Almost the only redeeming feature of the Jocko Junior is that for better or worse, it was the first racing car teenager Sam Posey ever drove. After two years off, in 1963 Jim Haynes won another National Championship, this time in Formula Junior, with a Lotus 18 and Lotus 20.

The next year, he bought an early Formula Vee, as well as gained control of Lime Rock Park. Haynes was President, Treasurer and Track Manager. For over two decades, Jim ran Lime Rock with grace and wit and was a major force in American racing. He was an early supporter of his friend John Bishop's IMSA races and a leader in restoring and racing vintage cars.

Harry Theodoracopulos bought Lime Rock Park in 1969, but Jim Haynes stayed on as track manager for another decade and a half, leaving to work on special projects for the SCCA. Today, Jim is President of Road America, the famous rural Wisconsin race track which has much of the bucolic ambience of Lime Rock.

Jim Haynes generously helped any number of young racing drivers, including Sam Posey and Rickie Bell, and always made coming to Lime Rock fun and exciting. For over thirty years, he's been a terrific spokesman for road racing and one of the true leaders of American motorsports. ■

(Left) Jim Haynes presents a bottle of champagne to Holger Ahl, winner of the Formula Ford Consolation Race at the Kendall Cup National in 1974. It was Ahl's first race win, ever. There wasn't supposed to be a Consi, there was no trophy, so Haynes came up with the champagne to make a grateful amateur racer even happier.
(Below) Haynes in a vintage race with his restored Fitch-Whitmore Special, 1976.

1964

April 25: LISCA. Neophyte racer Hans Rocke finished third in his class in an MGA.

May 2: SCCA New England Region. Among the new names, Peter G. Sachs in an F-modified Brabham. Charlie Gibson moved up to a Lotus 23. According to the program, James Edward Haynes had obtained control of the Lime Rock Corporation, with directors Mr. Haynes, John Fitch, Newton Davis and William Wilbourne. At the same time, the track offices were moved to a new building on Route 112. On the track, Bob Grossman won the big production race with his Ferrari GTO, lapping the field. Track record holder George Alderman did a 360 in the downhill, kept his foot in it and won his class anyway.

June 13: SCCA Northern New Jersey Region. Torrential rains delayed the day. The best race of the clearing afternoon was said to be the small-bore modified bash, won by Michael Goth in a Lotus 23 over Peter Sachs in a Brabham. Unusual enough to elicit comment at the time, there were two women, Janet Guthrie in her Xk-140 and Sylvia Zalla in her TR-4, in one race.

July 3-4: LISCA and Stock Car 100-miler. First there were four LISCA races, won by Tom Stafford in a Lotus 7, Gerry Sagerman in a TVR, Gaston Andrey in a Lotus-Alfa and Tom O'Brien in a Ferrari GT. There was also a vintage exhibition race started on handicap, one of the first vintage races in New England. Then they wheeled out 17 late model stock cars for the 100 mile Grand American Championship. Roy Haliquist beat Jim Reed and Bob Devine, all in Ford Galaxies. This race made Lime Rock Park one of the first road racing circuits in America to host late model stocks.

July 11: SCCA New England Region.

July 25: SCCA New York Region.

August 15: Lime Rock Park Championship Trophy. Another Lime Rock innovation: they paid expense money to amateur drivers and real cash prizes ($1250 for the open-wheel event). There was also a vintage race and a race for the hot new class of Formula Vee. The most startling finish of the day was in the big-bore modified race; Gordon MacKenzie (C-type) and Dick Stoultz (Lister-Jag) were leading as they came down the hill, when Joe Grimaldi popped out of nowhere with his D-type powered by a fuel-injected Chevy V-8 and out-dragged both Jaguars to the finish line. Most of the serious racers then drove overnight to Thompson, where they allow Sunday racing, and competed in an SCCA New York Regional.

August 29: SCCA Northern New Jersey Region. Bob Sharp won G-production in a Datsun, Hal Keck won A-production in a Cobra.

September 26: SCCA Northern New Jersey Region.

October 17: SCCA New England Region National. Peter Sachs lowered the outright track record to 1:01.8 with his Brabham-Ford.

October 24: SCCA New York Region.

Rock Corporation. As it turned out, Jim stayed at Lime Rock in that capacity for over two decades.

Among Haynes' immediate changes were to move the track offices to the small house they still occupy on Route 112. Previously, the offices had been in the base of the Timing Tower in the room that today is used as a Skip Barber School classroom during the week and for the timing and scoring computers on weekends.

Other overdue improvements included higher embankments along Turn Four and No Name Straight, red-yellow-green signal lights at the flag stations, better radio communications for the corner workers, extensive new fences around the spectator areas and new buildings on top of the infield hill and in the paddock. Haynes added another ambulance, a Datsun Patrol utility vehicle and a Datsun 410 pace car that was one of the first Japanese cars in all of New England.

Not everything was totally business. Among the nice little conveniences at Lime Rock in the mid-Sixties were Tuesday practice days every week for a fee of $5 per car—unmuffled practice, at that—and free practice on every Friday preceeding a Saturday spectator race. The gentlemen racers may have been gone, but a nice spirit of low-key congeniality still pervaded Lime Rock Park.

It seems almost incredible to us today, but before 1964, there weren't even programs at the Lime Rock races. Instead, events were covered by a regional publication called *Haybale*, that was sold at all the tracks in the Northeast. In 1964, Jim Haynes started publishing his own programs for each race, a revolutionary concept at the time.

Within a year, the Lime Rock program was taken over by well-known artist/photographer John Peckham and turned into a first-rate publication reminiscent of *Car and Driver* in both its innovative design and editorial irreverence. Peckham and his staff covered not only racing at Lime Rock, but everything from the Sebring 12-Hour to the annual hillclimb at Mt. Equinox.

To an extent not seen before nor since, John Peckham and Jim Haynes made Lime Rock Park the "Road Racing Center of the East," in the sense that not only was it was the premier road circuit and the "home track" for many drivers and clubs, but also everyone's source of racing news.

Eventually, Tony Muldoon and John Peckham turned the Lime Rock program into *Racing* magazine, which, like the old Haybale, was sold at Bryar and Thompson as well as Lime Rock. Starting in 1972, Joe Corbett and Greg Rickes ran *Racing* magazine; Corbett for some seventeen years, until 1988. Rickes and Corbett reported on lots of changes in both motorsports and Lime Rock Park over those decades.

(Below) A rare collectible; 1967 artwork by Sam Posey of his own Autodynamics Caldwell Can-Am car. (Bottom) Lime Rock program artist/editor/photographer John Peckham at the C/D Challenge in 1972.

The Rise of Professionalism

A far more widespread change than new management and a nicely designed program at Lime Rock was the increasing professionalism that permeated every level of road racing in the Sixties. At Lime Rock, Jim Haynes

(Continued on page 77)

GRAND AMERICAN SEDANS

Discounting the prewar Elgin Road Races and NASCAR racing on the beach at Daytona, the first time stock cars raced on a road circuit was at Linden, New Jersey in 1956. The first stock cars seen on a permanent road circuit were at Riverside over Memorial Day weekend, 1958. Eddie Gray won.

By 1963, there were eleven different races held around the country on road circuits for either stock cars, USAC sprint cars or midgets. Dan Gurney became almost a one-man stock car show, dominating the Riverside NASCAR race year after year. He even brought a Chevrolet Impala to Silverstone in 1961 to race in the British Saloon Car Championship, followed in 1962 by local Lime Rock star Peter Sachs and Charles Kelsey who raced Chevy IIs in England. In 1963, Jack Sears absolutely ran away with the British championship driving a Ford Galaxie. Dan Gurney, Jack Brabham and John Sprinzel also raced Galaxies in Britain in 1963.

If road racing a stock car was okay with a Formula I world champion, who could argue? And so for the July 4, 1964 weekend at Lime Rock, the events were the expected LISCA sports car races and an unexpected 100 mile Grand American race for stock cars. These were basically NASCAR stockers; local star Dick Dixon's year-old '63 Galaxie was the same machine Freddie Lorenzen had campaigned the year before into third in the NASCAR Grand National series. At the time, this Galaxie was the money-winningest car in racing history.

The United's Grand American Sedans were NASCAR's Grand National sedans. "Grand Prix" sedans had a minimum 116 inch wheelbase and maximum 427 cubic inch engine; "Super Sport" sedans used a maximum 365 cubic inch engine. A single four-barrel carburetor was the maximum limit, with the famous Ford "high-riser" and Chrysler Hemi V-8s specifically prohibited.

At the time, stock car sanctioning organizations were far ahead of sports car groups in terms of safety equipment. For example, rollcages—not just rollbars—were required. The cages had to have side intrusion bars along the doors and the doors themselves had to be permanently plated and bolted shut. Seat belts and catch tanks were mandatory, along with removal of the headlights, side windows and interior upholstery. Wheels were limited to 10 inch rim width, running 10.5 inch wide racing slicks. All this in 1964!

At Lime Rock, 1963 Grand American Champion Roy Hallquist from Bridgeport, CT started on the pole and won overall, driving a '64 Ford Galaxie sponsored by Loh Ford in Bridgeport but owned, amazingly enough, by Briggs Cunningham.

Late Model Stock Cars came back to Lime Rock on August 14, 1965 but it was really too tight and twisty for guys more used to banging fenders at Riverside Park and West Haven. They never returned...the nearest thing was Kelly American in the Seventies. But imagine if it had caught on! ■

(Top) Neither the first nor the last to get caught out by The Hook, the driver of this Super Sport '57 Ford sponsored by Chet's Late Model Wrecks just dirt-tracked it through.
(Lower left) Roy Hallquist from nearby Bridgeport, in his 1964 Ford Galaxie during a pit stop. The car was sponsored by Loh Ford of Bridgeport, but actually owned by Briggs Cunningham. Hallquist won the Lime Rock 100 mile Grand American Championship race in 1964.
(Below) Teammates in The Esses. Dick Dixon's #45 is the 1963 Galaxie driven by Freddie Lorenzen in NASCAR the previous year.

1965

April 17: SCCA New York Regional. Only 89 cars came out to open the season. Among the winners were Phil Groggins (Lotus Elan), George Oulton (Mini Cooper), J.D. Igleheart (Bobsy-OSCA), Ron Catalano (Fiat), Don Goodrich (Datsun SPL-310), Gerry Sagerman (TVR), Newt Davis (Lotus 30), Sy Kabak (Lotus 20) and Roger Barr (Autodynamics Vee).

May 1: SCCA New England Regional. The high excitement was Newt Davis setting his Lotus 30 on fire by forgetting to replace the oil dipstick before he went out to race. Typically, he calmly pulled to the side of the front straight, borrowed a fire extinquisher, ripped off the engine cover and put the fire out, all without exiting the car. Lucky for him, the flames never reached the fuel tank mounted between the engine and driver. Mike Goth won the race in his Cobra.

June 5: LISCA and ARDC Midgets.

June 19: SCCA Northern New Jersey Regional. Highpoint of the day was George Alderman in the Cooper-Alfa, who lapped just .1 off Peter Sachs' outright track record.

July 5: SCCA Northern New Jersey Region National. An interesting schedule for the weekend. Friday and Saturday, practice and qualifying, free beer party and steak dinner ($3.00). Sunday, "Continuous Shindig with 2 Combos" at Stagecoach Hill Inn. When you got tired of dancing, motorcycle races in Millerton, NY, neatly seguing into Stock Car races at Lebanon Valley Speedway. Those who were still mobile had to beat it back to Lime Rock for the first race at 10:00 am Monday. After the races, there was a fair in Lakeville, with fireworks. And they say racers don't know how to have a good time. The race program cover featured front and side drawings of a Chevy-powered Ferrari Monza, Chaparral II, McLaren-Elva and Ford GT-40. Only the Ferrari-Chev showed up for the Nationals, in the hands of Oscar Koveleski. As it turned out, the biggest crowd in Lime Rock history showed up for Monday's racing. Among the highpoints: Bob Sharp beat Dick Gilmartin and Phil Haloff in G-production, Brian Fuerstenau beat Art Riley and Al Costner in F-production, Candy DaMota beat Gaston Andrey in Formula B, Jim Haynes beat Floyd Aaskov in Quantum Saabs, Skip Scott beat Hal Keck and Mike Goth, all in Cobras, Bob Johnson beat Mark Donohue in Shelby Mustangs, Lew Kerr beat Ron Grable in E-modified.

July 24: SCCA New England Regional. George Alderman lowered the outright track record to 1:01.6 in his Cooper-Alfa.

August 14: BSCOA and Stock Car 100-miler. Jocko Maggiacomo—one of the rare personalities of that era accepted by both "circle burners" and "tea baggers," won the stock car race in a Ford.

August 28: SCCA Northern New Jersey Regional.

October 16: SCCA New York Regional.

MIDGETS

The famous Formula Libre race of 1959 was so successful that many circuits—including Lime Rock—tried to hold similar "run-whatcha-brung" events in the early-Sixties. For a variety of reasons, none of them succeeded. But what did succeed was racing oval track cars on road circuits, not all mixed up with sports cars and GP machines, but in their own races.

Riverside in California ran full-size sprint cars, but the smaller midgets were more suited to Lime Rock's tight twistiness. The most noteworthy event was in 1963 at Lime Rock and mixed American Racing Drivers Club midgets and SCCA Formula Juniors. Young Mark Donohue drove an Offy-powered Cooper Formula Junior that Kenny Brenn built as a mid-engine midget. Donohue in the "Brenn Pusher" ran rings around everyone, including the winner of one of the heat races, another young kid named Mario Andretti.

Two years later, on June 5, 1965, in the midst of a typical LISCA Saturday of sports car and formula races, the American Racing Drivers Club competed in what was called a "Monza-type race." This meant they ran three 10 lap heats; the starting order for the second and third heats was the reverse order of the finish in the previous heat. The overall winner, Russ Klar—who had also been in the 1959 Formula Libre race—received the Bill Randall Trophy, named in honor of a midget driver killed at Lime Rock in the 1963 event.

Among the stars was Len Duncan, who was a six-time ARDC season champion in the same front-engine Kenny Brenn midget driven by Rodger Ward in the 1959 Formula Libre race. Midgets ran again in 1972, on the IMSA GT weekend. ■

(Above) ARDC midget, complete with wing, during the Lime Rock event of 1972.
(Left) Finish of the first heat of the ARDC midget race at Lime Rock on June 5, 1965.

1966

June 11: SCCA Northern New Jersey Regional. Among the new names; Shirley VanKleeck in a Sprite, Al Cosentino in a Fiat-Abarth, Alan Bolte in a Sunbeam Alpine, Mike Rothschild in a Beach Formula Vee, Fred Stevenson in a Lotus Formula C. Among the new cars, Brandt Griffing had his homebuilt D-modified finally completed, Jim Haynes tried a Wasserman Formula C and Bob Colombosian a Lotus Elan. Highlights of the weekend: the continuing Bob Akroyd-Don Ward "feud" in H-production. Akroyd won this time. In G-production, Bob Sharp came in to change a fluffy sparkplug on lap one, restarted last and got all the way back up to second, 2 seconds behind winner Phil Haloff. John Peckam, editor of the Lime Rock programs, collected all the class lap records at Lime Rock for the first time. As of June 1966, they were:

AP: Posey, Porsche 904, 1:04.0
BP: Claydon, Lotus Elan, 1:04.7
CP: Stafford, Lotus Super 7, 1:09.5
DP: Howe, Jaguar, 1:08.4
EP: Opert, Elva Courier, 1:09.5
FP: Sharp, Datsun, 1:09.0
GP: Sharp, Datsun, 1:08.6
HP: Akroyd, Sprite, 1:12.0
Sedan: O'Neill, Lotus-Cortina, 1:11.8
Track Record and FB: Alderman, Cooper-Alfa, 1:01.6
FC: Donohue, Lotus 20, 1:02.5
FV: Barr, Autodynamics, 1:09.0
CM: Brown, Malibu Special, 1:03.0
DM: O'Brien, Ferrari Dino, 1:02.0
EM: Lake, Elva-BMW, 1:03.6
FM: Sachs, Brabham-Ford, 1:01.8
GM: Gibson, Lotus-Ford, 1:06.6
HM: Igleheart, Bobsy, 1:08.7

July 2-4: SCCA Northern New Jersey Region National. Once again, not only were there races on Saturday and Monday, but a "continuous shindig with combo" at Stagecoach Hill and stock car races at Lebanon Valley Speedway on Sunday. A new innovation was color-coded outfits for all the race personnel. Tech inspectors wore blue and white striped shirts and blue pith helmets, stewards had red and white striped shirts and red helmets, race chairmen and starters were in black and white stripes with white helmets, etc. In the Formula Vee race alone, new drivers included Ray Caldwell, Bill Scott, Sam Posey and Alex Dearborn. Sam also had a Porsche 904 for big-bore production, while Bob Leitzinger drove a Lotus Elan. Chuck Kirkbride brought his Formula A Lotus, Buzz Marcus a TR-4, Lake Underwood a Porsche and Tony Adamowicz a Lotus-Cortina. In the big sports/racing race, the most exciting cars were George Wintersteen's Corvette Grand Sport and Woody Young's Lola T-70 driven by Bob Bucher. Back in the pack was a young kid named Peter Gregg who came all the way from Jacksonville with his E-modified Porsche. The big-bore modified race was called the Walt Hansgen Memorial; Hansgen had just been killed at LeMans.

August 27: SCCA Northern New Jersey Regional.

September 24: SCCA Northern New Jersey Regional.

October 15: SCCA New York Regional.

followed John Fitch's lead by holding not only amateur sports car races, but professional events for midgets, stock cars and even, on one memorable occasion in 1968, a promotional demonstration before the Trans-Am in which a front-engine fuelie rail used the Main Straight as a drag strip, heading "backwards" from east to west with the start line in the run-off road of Turn One and the shut-down area in The Uphill!

Awakening SCCA

In a larger sense, the story of Lime Rock in the Sixties is the story of the SCCA's realization that whether or not the SCCA went along, road racing was going to split into two groups; those who raced for fun and those who raced for money. If the SCCA wanted to keep its self-proclaimed position as arbiter of American road racing, it was going to have to unbend a little and get involved with professional racing, like it or not. Otherwise, the sport was simply going to go roaring on past their cute Colonial ivory tower in upscale Westport.

Now back in 1958, John Fitch had been one of the leaders in bringing professional road racing to America. He'd gotten USAC to sanction everything from Formula Libre to big-bore sports car races. Some events were more successful than others, but the very first all-professional road race in America was held September 7, 1958 at Lime Rock Park. George Constantine won, driving his first race in Elisha Walker's Aston Martin DBR-2. There were only seven other finishers, the best being Bruce Kessler in a Ferrari. Not to put too fine a point on it, the first professional road race in America was an aesthetic and financial disaster.

True success came with the series of USAC races that started with the *Los Angeles Times-Mirror* Grand Prix at Riverside International Raceway in October of 1958. The *LA Times* is owned by wealthy car nut Otis Chandler and his family, so the *Times* Grand Prix was better promoted and more thoroughly covered than any race in the country. USAC and Riverside were also able to get approval from the Federation International de l'Automobile to call it an FIA race like Sebring or LeMans. This meant they could have European stars like Stirling Moss, Jack Brabham and the rest without the SCCA being able to say a word. It was a monumental coup.

Within two years, the *Times* Grand Prix had blossomed into an entire fall series of races at not only Riverside but Kent, Washington, Laguna Seca, California and Mosport, Ontario. These races cleverly bracketed the United States Grand Prix, which was always at the end of the season—run in 1959 at Sebring, then at Riverside in 1960 and after that at Watkins Glen. All the European stars competed in the lucrative Fall Series because they were already in North America, anyway.

From the European point of view, it was easy money. Cars like the Lotus 19, Cooper Monaco and Cooper "King Cobra"—not to mention Roger Penske's controversial modified Cooper Formula I car—were essentially designed just for this series. There was so much prize money involved it was worth it.

USRRC

In 1963, the SCCA struck back with a professional road racing series of its own. This was the United States Road Racing Championship, an eight-race series for the same sports/racing cars that had been competing in the USAC events. The USRRC was incredibly successful.

At the 1964 *L.A. Times* Grand Prix

(Above) A classic John Peckham drawing of the Surtees/Bandini Ferrari 330 P that finished third at Sebring in 1964 graces the cover of a Lime Rock program from 1965. (Overleaf) The 5.0-liter Chevy V-8 from David Hobb's 1972 Formula 5000 Lola.

at Riverside, for example, winner Parnelli Jones beat Roger Penske, Jimmy Clark, Dan Gurney, Bruce McLaren, Bob Bondurant and A.J. Foyt, among others. P.J. won $10,000 and a new Pontiac Grand Prix with a sticker price of $3400. At the time, these riches made the *Times* GP the best-paying road race not just in America, but in the entire world.

What does this have to do with Lime Rock? Well, in 1964 and '65, when everything was in great confusion, Jim Haynes took the opportunity to sanction his own professional events using alternate groups. The Long Island Sports Car Association sanctioned some offbeat weekends at Lime Rock, including stock car races that were joint ventures with the United Stock Car Racing Club and midgets that came from the American Racing Drivers Club and the Northeast Midget Association. This was a long way from the elite SCCA. Of course, this also didn't endear Jim Haynes to the SCCA. When the SCCA expanded its professional racing in 1966, Lime Rock Park was not included the first year.

The Revolution of 1966

In 1966, the SCCA came up with the three best ideas in racing. Taken together, the broad concept was that they would sanction three professional road racing series, one for sports cars, one for open-wheelers and one for sedans. The original concept was that all three would be based around the same 5.0-liter production V-8 engines, very much the way Formula I, IMSA GTP and FIA Group C share 3.5-liter engines today. They would also be run under FIA international rules, so drivers and teams from all over the world could compete. As it turned out, that's not exactly what happened, but pretty close.

Can-Am

The first of these series was the legendary Canadian-American Challenge Cup, run to the rules of FIA Group 7, Appendix J. From one point of view, the Can-Am was the best road racing series ever, an absolutely mind-boggling display of unlimited horsepower in some of the prettiest sports/racing cars ever designed.

On the other hand—and this is heresy, I know—the Can-Am was also one of the most boring race series in the world. After John Surtees won the championship in a Lola in 1966, McLarens won 37 of the 43 Can-Am races run from 1967 through 1971, after which George Follmer and Mark Donohue each ran away from the competition to win championships for Porsche. In terms of competition,

(Continued on page 82)

SAM POSEY

What can you say about Sam? He's almost intimidatingly bright, extremely analytical and certainly the most garrulous fellow I've ever met. If ever there had to be someone to fill up six hours of dead air on a rainy Indy 500 qualifying weekend, Sam is the one you'd want behind your microphone.

He also has the credentials to pontificate about racing. Sam inherited the means to go racing in style, it's true. He grew up only 5 miles from Lime Rock; as a teenager, he had John Fitch for a private racing tutor. His first race was in Formula Vee at Lime Rock in 1965; the next year he raced at LeMans, the next year he won the Sebring Trans-Am race, did the first lap of Lime Rock under 1 minute and finished third in the USRRC in a McLaren sponsored by Bob Akin's Hudson Wire. Nobody has ever rocketed to international prominence so quickly.

By 1968, Sam was running in the Trans-Am and Can-Am, in 1969 he won the Lime Rock Trans-Am and finished third in the Continental Championship for F-5000. He was also the only American to finish at Sebring, Daytona and LeMans, driving Ferraris for NART. By 1970, he was good enough to be fourth overall at Daytona and fifth overall at LeMans. He also had his own factory-backed Autodynamics Dodge Challenger in the Trans-Am, though the car was never a winner.

In 1971, Sam was second to David Hobbs in the L&M F-5000 series, fifth at both the Indy 500 and Pocono 500. He and Tony Adamowicz were third at LeMans; Sam set a new lap record in their Ferrari 512M. The next year he was second in the F-5000 series again and even got to drive a Surtees in the U.S. Grand Prix. After Mark Donohue was killed, Sam gave up open-wheelers, but he won Sebring with Brian Redman and Hans Stuckwith a BMW in 1975, then ran two seasons of IMSA GTU in a Bob Sharp Datsun, setting a Lime Rock lap record in 1977 and winning in 1978.

Of course, being Sam, at the same time he was also painting, designing new buildings for Lime Rock Park, writing for *Road & Track* and, since 1974, announcing the Indy 500. Some of Sam's wealthy Connecticut neighbors are proud of the fact that they've never had to *do* anything. Sam has used his wealth as a wedge to open doors so that he has been able to do *everything*. That has made all the difference. ■

(Above) After winning the Lime Rock IMSA race in 1978, Sam is more interested in a can of his sponsor's brew than the attentions of Lynn Griffis, Miss Camel GT. (Left) 1974: Sam has drawn a map of Lime Rock Park in the dirt and he's explaining to former track record holder Jim Haynes the quick away around!

1967

April 22: SCCA New York Regional. Due to ever-increasing insurance costs, the entry fee was reluctantly raised to $25. Among the inexperienced racers near the beginnings of their careers were Beck Beckwith in an Alfa, Tom Davey in Formula Vee, Tony Carbone in an MGA and Burge Hulett in a Lotus 23. Thanks to rapidly advancing tire technology and mid-engine sports/racers, nearly all the lap records had been lowered dramatically since the previous spring. The old "modified" classes were now more accurately called "sports/racer." As of April, 1967, these were the Lime Rock records:

AP: Lowther, Cobra, 1:04.0
BP: Claydon, Lotus Elan, 1:04.0
CP: Steubesand, Lotus Super 7, 1:07.8
DP: Howe, Jaguar, 1:08.4
EP: Frey, Porsche, 1:08.5
FP: Sharp, Datsun, 1:09.0
GP: Sharp, Datsun, 1:08.3
HP: Smith, Sprite, 1:11.7
Sedan: Lake, Mustang, 1:08.2
FB: Morin, Brabham, 1:01.5
FC: Donohue, Lotus 20, 1:02.5
FV: Barr, Autodynamics, 1:09.0
CS/R and track record: DaMota, McLaren, 1:00.5
DS/R: O'Brien, Ferrari Dino, 1:02.0
ES/R: Ashplant, Elva-BMW, 1:03.3
FS/R: Sachs, Brabham-Ford, 1:01.8
GS/R: Heald, Elva, 1:05.5
HS/R: Igleheart, Bobsy, 1:08.7

May 13: SCCA Mo-Hud Regional. More rookie drivers, including Charlie Fox in a Sprite and Hap Farnsworth in a Cooper-Ford.

May 29-30: SCCA Northern New Jersey Region Trans-Am Championship. Lime Rock's first Trans-Am was won by Peter Revson in a Mercury Cougar ahead of Mark Donohue's Camaro.

July 4: SCCA Northern New Jersey Region National. For years, the "one minute barrier" at Lime Rock had seemed as insurmountable as the sound barrier had during World War II. And when it finally fell, it fell just as easily. Sam Posey came to this lowly National with his USRRC McLaren-Chevy—sponsored by Hudson Wire and prepared by Ray Caldwell and Colin Day—for the express purpose of breaking the track record. He even wrote an article about it in the track program, sort of like Babe Ruth pointing out the spot where the next pitch would clear the fence. Sam won his National race—records had to be set in a race, not in practice—and left the outright track record at 58.0 seconds. Mr. and Mrs. Stephen Mayer, owners of MEM, presented him with the English Leather Cup. Later that summer, Skip Barber, driving a similar car, lowered the one lap record to 56.3.

July 29: SCCA New York Regional.

August 26: SCCA Northern New Jersey Regional.

September 2-4: SCCA New England Region National.

September 23: SCCA Northern New Jersey Regional.

GOODYEAR

(Previous page) Murph Mayberry in his Lime Rock shop with the Lolas of David Hobbs and Brett Lunger, 1972. (Above) Formula 5000 stars Dick Smothers and David Hobbs. (Right) Rising star Ron Grable, in 1970. He's Editor-at-Large of Motor Trend, *today.*

the fondly-remembered Can-Am was really a snooze.

Formula 5000

The second of the SCCA's professional series was for formula cars. Now you will remember that the *Sports Car* Club of America had traditionally been lukewarm towards formula cars. Maybe it was because gentlemen raced sports cars and professional drivers were hired to compete in Grands Prix. Whatever, right up until modern times, a typical SCCA Regional or National weekend would have eight or nine races for sports cars, and one race in which all the formula cars, from Formula Vee to Formula I, were thrown in together. If you didn't have fenders, it was enough to give you a complex.

In 1968, Greg Rickes—who has been announcing races at Lime Rock for over two decades now and who is also a penetrating observer of the road racing scene—wrote a thoughtful little essay in *Racing* magazine. In it, Rickes pointed out, "Motor racing attains its purest state with open-wheel racing. The formula car is what racing is all about...it is the car in all its essence, unmistakeably designed for the ultimate in performance without compromise."

In 1966, as part of the huge reorganization, the SCCA took all its various open-wheel classes and consolidated them into three formulas. Formula C was for cars that weighed at least 750 lbs. with maximum engine size of 1100cc. In other words, leftover Formula Juniors. Formula B was for cars that weighed at least 848 lbs. and used production-based engines between 1100cc and 1600cc. Essentially, this meant you could fit your old Lotus Formula Junior or 1.5-liter Formula I car with a 1600cc double-overhead cam engine from an Alfa-Romeo GTA or the ubiquitous Cosworth-Ford Twin Cam. This was a combination that at Lime Rock, at least, was capable of seizing the outright track record.

And then there was Formula A. Formula A was essentially for Formula I cars from any previous or current Grand Prix formula—1.5-liter, 2.5-liter, 3.0-liter—that used "pure racing" engines. The cars had to weigh at least 1105 lbs. and 1.5-liter cars could be supercharged. There was also a second type of Formula A car, essentially a 3.0-liter Grand Prix machine fitted with a 5.0-liter "stock-block" from a series produced American passenger car. These cars had to weigh a minimum of 1250 lbs.

Interestingly enough, the rules for Formula 5000 not only encompassed the new 3.0-liter Formula I that started in 1966 but were also very close to the rules for the Indianapolis 500 and the USAC National Championship. The basic USAC formula was for racing engines 2.8-liters supercharged or 4.2-liters unsupercharged in cars that could weigh as little as 1350 lbs. in 1966. In 1966, USAC also raised the stock-block limit to 5.0-liters (5.25-liters starting in 1969).

In practice, what all these new rules meant was that a racing car constructor like McLaren, Eagle or Lola could design and build one type of car that with only minor modifications could be successfully raced in Formula I, USAC and Formula 5000. Engine builders benefitted from rules rationality, too. The Eagle with which

(Left) Lime Rock Schaefer Grand Prix, 1969: George Eaton #98 leads eventual winner Peter Gethin #57 through the Left-Hander. Behind them are Bobby Brown #3, John Cannon #2, Lothar Motschenbacker #11, Dick Simon #66, George Wintersteen #12 and Sam Posey #1. In its heyday, Formula 5000 had more famous names than Formula I. (Below) David Hobbs' 1971 McLaren—owned by Carl Hogan, maintained by Murph Mayberry, sponsored by The Hot Comb—passes the Lime Rock concession stand.

(Right) Chris Economaki interviewing Schaefer Grand Prix winner Peter Gethin. 1969.
(Far right) George Wintersteen, 1969.
(Below) David Hobbs on his way to lowering the Lime Rock track record to 50.8 with his Surtees TS-5A, 1970.
(Opposite page) Long-suffering crews try to sort things out after their drivers have made a muck of things.

Dan Gurney finished second at Indy in both 1968 and 1969 was essentially a Formula 5000 Eagle whose Trans-Am Ford V-8 had been fitted with new cylinder heads designed by Harry Weslake, who of course, had also designed the 3.0-liter V-12 that had powered Gurney's similar Formula I Eagle in 1967. Thanks to cooperation between the FIA, SCCA and USAC, racing all over the world was improved.

After the SCCA's 1966 shake-up, it took a while to put everything into place. The professional SCCA Grand Prix Championship series didn't really get going until 1967, and they didn't figure out that you could easily slip a Trans-Am stock-block into a Formula I car until the 1968 season. So Formula 5000 as we remember it didn't really begin until 1968. It was worth the wait.

Continental

The SCCA Grand Prix Formula A Championship started in 1968. California dentist Lou Sell, driving an Eagle/Chevy sponsored by the Smothers Brothers, was the big gun that first year, winning the championship. Among his competitors were Jerry Hansen, Bobby Brown, Sam Posey, teammate Dick Smothers (who ran Formula B) and series runner-up George Wintersteen. Wintersteen did win the Lime Rock race, however, the only race out of eight that he won all season (though he did finish first overall at Donnybrooke by coming second in each of two heats).

This was Big Time professional road racing by the standards of the day. The Mayer family, which owned English Leather cologne, was the event sponsor at Lime Rock. Having an "event sponsor" was a pretty innovative concept in 1968. The total purse was a whopping $12,000, of which Wintersteen collected $2100. For the season, Sell won $13,700 and Wintersteen $12,050. Even in 1968, that was about a quarter of what they needed to run the series. Racing is expensive.

In 1969, Lime Rock ended up with

(Continued on page 88)

16
FEINSTEIN RACING

1968

April 27: SCCA New York Regional.

May 11: SCCA New England Regional.

May 30: SCCA Trans-Am Championship. Mark Donohue was on his way to a fairytale season, winning 10 of 13 Trans-Am races—8 in a row—and starting from the pole 9 times. Lime Rock was one of those races. Mark qualified ahead of Peter Revson's Javelin and the Mustangs of Jerry Titus and NASCAR star David Pearson, led into the first corner and continued to lead for 169 laps. The only real excitement was the under-2.0-liter battle of Horst Kwech and Tony Adamowicz, with Adamowicz winning after Kwech broke. The finishing order was Donohue, Titus, Revson, Adamowicz, then Bob Bailey-Jim Locke who inherited fifth in their Porsche after Malcolm Starr's Mustang ran out of gas within a few laps of the end. Mark drove overnight to St. Jovite and won the USRRC race the next afternoon. Incredible.

June 15: SCCA New England Regional.

July 6: SCCA New Jersey Region National. Lap records continued to fall faster than ever before, or since. You'll notice that there have also been some revisions, due to cars changing class, class changes and timing inaccuracies in earlier years. At the middle of the 1968 season, the official lap records were:

AP: Lowther, Cobra, 1:04.0
BP: Duclos, Corvette, 1:06.5
CP: Leitzinger, Lotus Elan, 1:05.1
DP: Sharp, Datsun, 1:06.7
EP: Bailey, Lotus Super 7, 1:06.5
FP: Sharp, Datsun, 1:07.6
GP: Adamowicz, Spitfire, 1:08.0
HP: Domizi, Sprite, 1:09.2
ASedan: Brown, Camaro, 1:06.4
BSedan: Adamowicz, Lotus-Cortina, 1:07.0
CSedan: Miller, Mini-Cooper, 1:09.9
DSedan: Johnson, NSU, 1:12.0
FA: Kirkbride, Lotus, 1:02.4
FB: Morin, Brabham, 1:01.5
FC: Donohue, Lotus 20, 1:02.5
FV: Barr, Autodynamics, 1:09.0
CS/R and track record: Posey, McLaren, 58.0
DS/R: O'Brien, Ferrari Dino, 1:02.0
ES/R: Ashplant, Elva-BMW, 1:03.3
FS/R: Sachs, Brabham-Ford, 1:01.8
GS/R: Heald, Elva, 1:05.5
HS/R: Igleheart, Bobsy, 1:08.7

July 20: SCCA Mo-Hud Region National.

August 24: SCCA Northern New Jersey Regional.

September 2: English Leather Grand Prix Championship. George Wintersteen (Eagle-Chevy) won from Bobby Brown (Lola-Chevy). Formula B class winner Fred Stevenson was an impressive fourth overall.

September 21: SCCA Northern New Jersey Regional.

October 12: SCCA New York Regional.

MURPH MAYBERRY

(Below) Bill "Murph" Mayberry in an unusually pensive mood, 1967.

Just as there are famous racing drivers, so too, there are famous racing mechanics...witness Lime Rock resident Bill "Murph" Mayberry. Murph has seen it all. As a teenager, he went to the races with Indy crew chief George Bignotti; when he got out of the service in the early-Fifties, he went to work for a Porsche dealer in Pennsylvania named Bob Holbert. Pretty soon, Murph was racing Porsches himself and maintaining Bob Holbert's cars.

In 1959, a young racer named Roger Penske bought Holbert's old 550 RS, so Murph started maintaining Penske's cars as well. Roger introduced him to another wealthy young racer, George Wintersteen. Murph Mayberry worked on all of Roger's cars, from the Porsche through the Corvette Grand Sports, on Wintersteen's cars from the Grand Sports through the Formula 5000 cars.

When Mark Donohue joined Penske, Murph was the one who maintained the Can-Am Lola and Donohue's F-5000 car. He also built the famous Camaros with which Penske Racing dominated the Trans-Am as well as the less-successful Penske/Kirk White Ferrari for endurance racing.

At some point in the late-Sixties, it all got to be too much. "We were trying to do too many things," says Murph. "We had race cars all over the place, but we never had enough time to get them right." He joined up as the crew chief on David Hobbs championship-winning F-5000 car in 1970 and 1971. In 1972, he moved the whole Hogan Racing operation across the street from Lime Rock, into the building that's now the Paradise Garage.

"We got a Chevy V-8 from Bartz, with a 180 degree crankshaft and we put it in David's Lola. Bartzie told us, 'That engine will only last 3 laps of Lime Rock.' That car had wheels that were 22 inches wide in the rear, wings, everything. The engine only lasted two and a half laps. But the second lap was a 45.9. Remember, that's twenty years ago. The SCCA outlawed 180 degree cranks after that."

When did he quit? "At Pocono in 1972, David's teammate Brett Lunger went out in the rain and crashed our third new car in two weeks. When he was 'on,' Brett was really good. When he wasn't concentrating, he crashed a lot of cars. I guess I was burned out. I called the team together in the Pocono pits next to this wrecked race car, and I said 'It's been great, fellas. I'm outa here.' I went back to Lime Rock, packed everything in a truck and shipped it to Carl Hogan in St. Louis."

Old racers never retire. Murph built some IMSA Porsches for Sam Posey and ran Indy with him, then discovered vintage racing. He has restored and raced dozens of vintage cars. Today, Murph Mayberry works in his one-man shop, a shop so clean you can see your face in the floors, restoring cars he likes for people he likes. The machines run the gamut from priceless vintage racers to somebody's old Chevy truck. And as he says, "If I want to head for Florida, there's nobody to yell 'Hey you, get back in here.'" ■

changed. Hobbs, Cannon and Wintersteen were the only front-rank drivers who hadn't moved on to something else, and the series-winning AAR Eagles were basically gone. The high-mounted wings were still legal for the American races, but outlawed for the Canadian ones. Everyone already knew that these dangerous high wings would be legislated away for 1971, so many cars were festooned with the lower, but not as effective "tea tray" wings used in Formula I and USAC since 1969.

The series was won in 1970 by John Cannon driving a Chevy-powered McLaren owned by Malcolm Starr and Carl Hogan. Other race winners in 1970 were Gus Hutchison with a 3.0-liter Brabham-Cosworth Formula I car, Ron Grable with a Lola-Chevy, George Follmer with a Lola-Ford, Mark Donohue with a Lola-Chevy and David Hobbs with a Surtees-Chevy. Compared to contemporary Can-Am races, Formula 5000 had great diversity in both cars and drivers, which made for terrific racing. The Continental Championship really was approaching world-class status, and drivers, sponsors and teams were quick to seize the opportunity.

The Great American Racing Series

The glory days of Formula 5000 were 1971, '72 and '73. L&M picked up the series sponsorship and the

(Above) A VIP ticket to the Lime Rock Trans-Am, 1971.
(Right) As entertainment before the 1968 Trans-Am, Jim Haynes got Dave Mott to run his front-engine dragster the wrong way up the Main Straight. According to Haynes, "He got so much bite, the thing did a wheelie and when it came down, it broke the frame!" This was just one of Haynes many promotional ideas. His favorite was when they flew model airplanes from the Straight. "One airplane got out of control and crashed right into Murph Mayberry's brand-new Lincoln Mark IV. Put a hell of a dent in the door," says Haynes with a chuckle.

series purse expanded dramatically to $305,000 or $35,000 per race. It was now lucrative enough to draw drivers from all over the world, at least some of whom came away with bigger reputations than they brought in. David Hobbs won the series in 1971, and as Lime Rock announcer Greg Rickes wrote at the time, "The Continental Championship has made David Hobbs, and David Hobbs has made the Continental."

In 1972, the competition was even stronger. L&M raised the race purse to $40,000 plus $10,000 just for qualifying. Among the name drivers who showed up at Lime Rock on September 4 were David Hobbs, of course, plus his teammate Brett Lunger, Horst Kwech, Skip Barber, Lothar Motschenbacher, George Follmer, Sam Posey, Rocky Moran, Horst Kroll, John Gunn, Graham McRae, Brian Redman, Bill Brack, Eppie Wietzes and Stuart Forbes-Robinson. Granted, it wasn't the Indy 500 or the Monaco Grand Prix, but the cars were as fast or faster than USAC and Formula I machines, and there are certainly some world-class drivers on this list.

How fast were they? David Hobbs ran half-a-dozen races in England that mixed Formula I and Formula 5000 cars in 1972, says his mechanic Murph Mayberry. "That's the most fun I've ever had. David put our 5000 car in the top ten Formula I cars. Those 5000 cars were *fast*."

As it turned out, New Zealander Graham McRae won the Tasman F-5000 Championship at home, then blew away Sam Posey, Brett Lunger and Brian Redman to win the L&M Championship as well. He also dominated the European F-5000 season and made enough money to buy Leda, the maker of his racers. Not everyone gets to win major championships with a car carrying his own name on the hood. Nobody could have had a more satisfying season.

In 1973, Jody Scheckter ungraciously grumbled his way into the L&M Championship, beating Brian Redman, Mark Donohue, Peter Gethin and David Hobbs, in that order. But while the championship continued, L&M money went elsewhere. As for Lime Rock, they traded their fading F-5000 race for a date with the new IMSA group, begun by disaffected SCCA head John Bishop. But that's another story.

Formula 5000 is one of those curious semi-

(Above) Famous timing and scoring guru Judy Stropus keeping track of the AMC Racing Team Javelins during the 1971 Lime Rock Trans-Am. (Left) Peter Revson and Carroll Shelby in the Lime Rock paddock prior to the 1969 Trans-Am. As it turned out, Revson qualified for the Indy 500 run the same day as the Trans-Am. He finished fifth. Sam Posey drove Peter's Shelby at Lime Rock, and won.

failures that pop up time and again in sports. There was money in the series, there were famous drivers, the competition was absolutely ferocious, the cars were fast and loud and exciting. For a while there, when it seemed like racing would be "The Sport of the Seventies" as Brock Yates kept promising in the pages of *Car and Driver,* Formula 5000 seemed like it might be The Great American Racing Series. Too bad.

Trans-Am

In 1966, when the SCCA decided to start its triumverate of professional racing series—the Can-Am, the Continental Pro-Formula and the Trans-Am—I'm sure they expected the first two to be the great success. Trans-Am started out as just a way to get the automobile manufacturers involved with road racing.

But it quickly became much more than that. Of all the road racing series that the SCCA has sanctioned, Trans-Am is the one everybody remembers. Do you associate Mark Donohue with winning Indy in 1972? Do you even remember who won Formula 5000 in 1974? But every road racing fan in this country over the age of twelve knows Mark Donohue drove a Camaro in the Trans-Am.

In 1966, the SCCA was trying to strengthen its ties with the FIA by making sure that its races were run for FIA approved cars. The FIA, of course, has a class to race almost anything with four wheels. Sedans come under FIA Appendix J, Group 2. At the time, there was already a highly-successful European Touring Car Championship, plus a British saloon car series. The big guns were

(Above) In the pages behind the original oil painting on the cover of the 1971 Trans-Am program there was a big story and advertisement about Steve McQueen's* LeMans*, now considered the best racing movie ever made. Can that really have been more than twenty years ago? (Right) Is it real, or is it...memories? This 1990 recreation of the 1970 Trans-Am race at Lime Rock is so authentic even Sam Posey couldn't tell the difference.

Jaguar, Lotus-Cortina and the Alfa Romeo GTAs.

For amateur events, the SCCA divided sedans into A/sedan (2.0 to 5.0-liter), B/sedan (1.3 to 2.0-liter), C/sedan (1.0 to 1.3-liter) and D/sedan (under 1.0-liter). But for their new professional series, they would run only two classes; under 2.0-liter and 2.0 to 5.0-liter.

The first year, these professional races were meant to be FIA endurance races between manufacturers, and required a mandatory driver change. Since the participants had to literally cross the country from Sebring to Riverside, the series was called the Trans-American Sedan Championship. And thus is immortality born.

Trans-Am 1966

Wobbly Dodge Darts, Plymouth Barracudas and Ford Mustangs were barely quicker than the Alfa Romeo GTAs and Ford-Cortinas despite their comparatively huge V-8s. Thanks to Shelby-prepared, factory-backed Mustangs which won the last two races at Green Valley, Texas and Riverside, California, Ford won the Over 2.0-liter Trans-Am. Lime Rock heros Gaston Andrey and Horst Kweck ran away with the small car championship driving Alfa Romeo GTAs.

So many 1966 Trans-Am drivers became famous later on. Bob Tullius

(Continued on page 97)

(Far Left) Jim Hall and Ed Leslie ran identical Chaparral Camaros in the 1970 Trans-Am. Neither won a race, though Vic Elford did win at Watkins Glen, driving the injured Hall's car.
(Left) In 1973, the Trans-Am rules changed to something between IMSA GT and NHRA Pro Stock. Camaros could run 460 cubic inch big blocks; Corvettes weighed under 3000 lbs. with 454 cubic inches. It was a great concept, but it was no longer the Trans-Am.
(Overleaf) The real Trans-Am: Mark Donohue in a Sunoco blue and yellow Camaro at Lime Rock, Memorial Day, 1968. Who could ever forget?

SUNOCO CAMARO
6

PENSKE-HILTON RACING

open rules had made the big cars dramatically faster than they'd been two years earlier.

Trans-Am 1971

All the other factories quit after 1970, so Penske and Donohue returned with the factory Javelins in 1971 and won eight of ten events. They blew off a bunch of independents, including George Follmer in his old Bud Moore Mustang. In an effort to open up the small car ranks, the SCCA raised the Under limit to 2.5-liters. That started the reign of John Morton in the famous Pete Brock BRE Datsun 510s, with Bob Sharp running similar cars on the East Coast. BRE dominated Under 2.5-liter as severely as Penske dominated Over 2.5-liter, winning the championship in both 1971 and '72.

This was the year of the "Great Deluge" at the Lime Rock Trans-Am. Bob Tullius, driving Herb Adam's '64 Pontiac Tempest, slipped through the field and got as high as second before the car quit. This was one of the best drives of the whole Trans-Am series, and one of the best races of Tullius' career. Not to mention, a textbook demonstration of car control in slippery conditions...who will ever forget that Tempest coming down the hill and sending up a rooster-tail seemingly three stories high. There never was a performance like it.

Trans-Am 1972 and Beyond

In 1972, there were no factory teams; George Follmer's Roy Woods Javelin beat the usual underfinanced independents, winning four of the seven races held. In 1973, the SCCA switched to FIA GT rules, which allowed Corvettes and Camaros up to 460 cubic inch big-blocks and 429 big-block Boss Mustangs. Even the Lime Rock Trans-Am program of 1973 called it "caricature racing."

This was just a pale imitation of the "real" Trans-Am, those five years when some of the best racers in history, driving some of the neatest race cars ever conceived on an unlimited budget, battled fender to fender to see who really was the best. They truly were five years of glory.

Lime Rock Central

Under the direction of Jim Haynes, Lime Rock Park was in the very center of these new professional racing series, Formula 5000 and Trans-Am, and at the very heart of the monumental changes that permanently altered road racing. Lime Rock managed this transition from gentlemen amateurs to highly-paid professionals with unusual grace, preserving the old ambience while accomodating the new intensity. Even when involved in an all-consuming struggle that might mean the survival of the team, it was still fun to come racing at Lime Rock. Take the fun out of professional racing, and you might as well be selling refrigerators...racing needs to be *fun*. At Lime Rock, it always seems to be.

4. Serious Amateurs

Today's Racing Quiz: Yes, this is Lime Rock Park. Yes, this is a Datsun 240Z. Yes, it belongs to Bob Sharp. Yes, it is about to win the C-production National on September 5, 1976. Who's driving? No, not Bob Sharp. Not Sam Posey. Not Paul Newman. Not Bruce MacInnes. Not Jim Fitzgerald. Would you believe Elliot Forbes-Robinson?

(Opposite page) Local Boys Make Good: Lime Rock regular Lew Kerr won the1963 ARRC at Riverside in this Formcar Formula Vee owned by another Connecticut resident, John Timken. The two won another National Championship in 1964 with a sports/ racing Brabham BT8. Kerr won every race he entered with the Brabham, including the ARRC at Daytona.

Lime Rock almost didn't make it into the Seventies. In the fall of 1968, a Connecticut legislator from Stamford got a state law passed to prevent noisy snowmobilers from riding across his backyard. So as not to discriminate, the law was written preventing the operation of unmuffled internal combustion engines in the State of Connecticut!

Lawyers for the Lime Rock Protective Association quickly got an injunction to stop racing at Lime Rock under the provisions of this new law; Jim Haynes had to post a non-refundable $10,000 bond in order to finish the 1968 Lime Rock season. Then he went looking for this legislator, who turned out to be a retired oval track midget racer who was surprised that his own anti-snowmobile measure was being used against Lime Rock. It still took Jim Haynes most of the winter to get the law repealed.

Rebirth

Simultaneously, the five-year agreement between Frank Vaill and John Norwood expired. "By the provisions of the agreement," explains Jim Haynes, "Mr. Norwood was required to sell the track property back to The Lime Rock Corporation at a specified price. At that point, I negotiated a sale, through the corporation, with wealthy racing enthusiast Harry Theodoracopulos. He obtained sole ownership of the track in the winter of 1969."

Thanks to some quick maneuvering, the track opened on schedule for the 1969 season. "Harry T" even won the first race with his Alfa Romeo GTA. Jim Haynes stayed on as the track manager for another fifteen years, and this purchase turned out to be a very workable arrangement for everyone, probably saving the track from becoming yet another housing development.

At the time, saving Lime Rock was a sound business decision; amateur racing was booming, professional racing was in the midst of a huge upswing, the crowds were large and growing and optimism filled the air along with the growl of Traco V-8s. Within three years, however, motor racing—not just at Lime Rock, but world wide—was under serious attack.

The "Energy Crisis" made targets of all sporting events because of the fuel burned by spectators driving to get there; motor racing was an even more obvious target because the competing cars so obviously burned precious hydrocarbons. By 1974, when otherwise responsible leaders were calling for a ban on all racing, buying a race track didn't seem like such a smart decision after all.

The recession of the Seventies and the second "Energy Crisis" of 1978-'79 made things seem even worse. But like everything else in life, this was just a cycle. By the early-Eighties, both racing and the economy were booming again and owning a picturesque race track less than 100 miles from Manhattan looked like a pretty smart thing to do. For Harry Theodoracopulos and Jim Haynes, however, the Seventies were a difficult period.

Period Charm

In general, Lime Rock Park was not that much different in the late-Sixties and early-Seventies than it had been in the Fifties...or than it is today. In detail, however, it was much more primitive than it is now. The pit lane was separated from the race track by only a single row of guard rail, and still was not paved. The rudimentary buildings were still the same ones that had been there since the track was opened. Even the ambulance was the famous '56 Cadillac that sat rooted to a spot next to Pit Out for decades; the track tow trucks were genuine antiques.

About the only major change was a new timing and scoring tower on the outfield side of the front straight to replace the primitive tower that was still over on the inside of the

short straight between The Hook and the Left-hander. This new location lasted just a few years. It was so close to the track, rooster tails from passing cars on a rainy race day would blow right into the building. The rainy 1971 Trans-Am was particularly poor in that regard and contributed to the building's quick removal. Few people remember that it was ever there, and nobody mourned its passing.

(Above) Winner Don Ward in his Cooper-BMC Formula C, 1968.

Races were started from a standstill rather than with a rolling start up until 1970. In those days, chief starter Jim Chipps literally stood on the race track with his flags, and in 1971 almost got cut down by Franz Weis in the long-forgotten Chaparral Formula 5000 car. Soon after, the existing starter's bridge was built from the timing tower out over the pit lane, raising the starter and his crew well above the danger zone.

Fitch Barriers

Another important safety device was engineered at Lime Rock during this period...John Fitch developed his famous "Fitch Barriers." During World War II, while he was stationed in Tunisia, John's squadron protected their bunks from strafing with sand-filled gas cans. Twenty years later, when he was looking for an impact attenuator to soften the ends of the guardrails at Lime Rock, John tried stacking up sand bags.

After nearly a decade of work, these crude sand bags evolved into the patented Fitch Inertial Barrier System, those ubiquitous yellow plastic barrels filled with sand stacked in front of highway hazards around the world. Virtually all of the testing was done at Lime Rock Park, and the first installation of these safety devices in the world was to protect drivers from the end of the guardrail at the entrance to the Lime Rock Pits. This was precisely the spot that John Fitch had wanted to protect when he started thinking about his barriers in the late-Fifties.

Turning Points

In the early postwar days, nearly all American road races were SCCA amateur events that counted toward the National Championship. Instead of depending on just one annual "runoff" race, the season championship in each class was decided on the maximum number of points each driver garnered during the season. Every SCCA race became, perforce, a "National."

Soon, there were so many races that only the most important ones carried National points; the others were called "Regionals" because they were run by the dozens of individual SCCA regions. By the late-Fifties, in other words, there was already a division between Nationals and Regionals, between quasi-professional and true amateur racing.

"Alternate" sanctioning bodies also put on amateur road races. In Southern California there was the famous Cal Club, almost totally autonomous of the SCCA. At Lime Rock, the inde-

(Continued on page 107)

HARRY THEODORACOPULOS

It made a great newspaper article. On May 1, 1969, Lime Rock held its first races since the track had been closed by a court order after Labor Day the previous year. In the meantime, the track had also been sold.

The new owner was a young Greek shipping heir named Harry Theodoracopulos. He drove an Alfa GTA in the first race of the 1969 season, the first race since the track had been closed, the first race since he'd bought the place. And he won! "New Race Track Owner Wins Lime Rock Race" was the headline. Some guys are like that; whatever they do is "news."

Harry T started competing in 1959, and won his class—third overall—in the 1960 Acropolis Rally. After returning to the States, he shared an Alfa with Sam Posey at Bridgehampton and a Porsche 904 at Daytona. These were some of Sam's first "professional" rides.

Harry went on to win the Northeast Division Championship in 1968 in B-sedan, then won both the Divison and the SCCA National Championship in 1969, driving a C-sedan Alfa. That led to similar Alfas in the Under 2.5-liter Trans-Am with his friend and teammate Horst Kwech.

In 1973, Harry and Horst switched to the all-conquering 2.9-liter Capris and subsequently to the even more dominant Chevrolet Monzas. Harry destroyed one of those against the Lime Rock Armco at the top of The Uphill. He retired from racing after the 1975 season. As Harry says, "Racing was always a weekend thing with me, like other people play golf."

In a sense, so was Lime Rock a "weekend thing." As Harry puts it, "I thought the track had good possibilities. I thought Jim Haynes was a good manager. I thought Butch and Donna did a good job. I knew some of the local people socially, and I tried to cement our relationships with the neighbors. Then I let Jim Haynes run the track.

"After I stopped racing, I didn't go there as much any more. When Skip Barber and his backers made me an offer, I accepted. One, I made a good profit and Two, I thought they would be good owners, involved with the track on a daily basis. They would make it grow and prosper. I must admit that I have not been back since I sold it, and I have not had anything to do with cars or racing. For me, Lime Rock is like an old mistress that I do not want to see anymore." ■

Lime Rock Park owner Harry Theodoracopulos in his famous blue "Lime Rock" Monza ran both IMSA and SCCA in the mid-Seventies. He destroyed this car in the Lime Rock Uphill during a practice session, getting spectacularly airborne in the process.

1970

Big news was a weekly two-day racing drivers school sponsored by Lotus Racing East and taught by Freddie Stevenson. The students used Lotus 61 Formula Fords.

May 9: Schaefer Trans-Am. This year the Lime Rock round of the Trans-Am was shifted from Memorial Day to avoid a conflict with the Indy 500, so all the series stars showed up. Mark Donohue qualified on the pole with a 58.8 in his Javelin, ahead of Dan Gurney's Barracuda. Donohue lost oil pressure halfway through, as did his Penske teammate Peter Revson and Mustang driver George Follmer. At the end, it was Parnelli Jones over a lap ahead of Ed Leslie's Chaparral Camaro, Sam Posey's Dodge Challenger and Jim Hall's Camaro. Jones won a whopping $3500. The under-2.0-liter race was run as a separate event this year, with Gaston Andrey and Horst Kwech 1-2 in Alfa-Romeo GTAs. Only one driver who competed in this Trans-Am is still actively racing; Lime Rock Expert and Camaro privateer Peter Schwartzott, who today races a Honda in IMSA Firestone Firehawk.

July 5: SCCA New York Region Amalie National: So many things happened simultaneously; an increase in sponsorship, an increase in professionalism, an increase in technology, and logically, a substantial decrease in lap times. At the start of the 1970 season, these were the lap records:

AP: Grossman, Cobra, 1:02.2
BP: McLean, Cobra, 1:03.9
CP: Pulver, Lotus, 1:03.4
DP: Sagerman, TVR, 1:05.3
EP: Frey, Porsche, 1:05.5
FP: Furstenau, Spitfire, 1:05.8
GP: Adamowicz, Spitfire, 1:08.0
HP: Seavey, Sprite, 1:08.2
ASedan: Posey, Mustang, 59.2
BSedan: Kwech, Alfa-Romeo, 1:03.4
CSedan: Theodoracopulos, Alfa-Romeo, 1:06.0
DSedan: Johnson, NSU, 1:12.1

FA and track record: de Adamich, Surtees, 52.1
FB: Stevenson, Lotus, 55.0
FC: Mitchell, Brabham, 58.5
FF: Barber, Autodynamics, 59.4
FV: Scott, Zink, 1:04.3
AS/R: Barber, McLaren, 56.3
BS/R: Mitchell, Brabham, 1:00.2
CS/R: Gibson, Lotus, 1:03.8
DS/R: Igleheart, Bobsy, 1:05.4

Among the winners that weekend were Bill Mitchell in his B-sports/racing Brabham, Bill Holbrook in C-sports/racing, John Paul in A-production and Tom Rizzo in B-production, both in Corvettes, Bob Sharp in a D-production Datsun, Bob Tullius in a C-production TR-6 and Eric Kerman in a D-sports/racing Arachnid.

September 7: L&M Grand Prix. David Hobbs not only won the Formula A race, he set a new Lime Rock race average speed record at 107.44 mph and a new track record at 50.8. Mike Eyerly won the Formula B race; Mike Rand won the B/C "Consi" in his Formula C car.

pendent Westchester Sports Car Club and Long Island Sports Car Association sanctioned frequent races for over a decade.

LISCA weekends were especially low-key, mostly because many of the big-name drivers who were chasing the SCCA Championship saw no reason to race when there were no points to win. In a very real sense, LISCA races served as the "Skippy School" of the late-Fifties and early-Sixties, introducing neophyte drivers to road racing in a congenial, non-threatening atmosphere.

At the same time, there was a third level of racing, true professional racing for cash prizes that had started with the USAC series in 1958. By the early-Sixties, most of the hot-dog professionals had been siphoned off into the USRRC, leaving Nationals and Regionals to the true amateurs. You might still see rising stars like Mark Donohue at a Lime Rock National in the mid-Sixties, but after 1966, once the serious money had come into the Trans-Am, Can-Am and Formula 5000 championships, professional drivers not only had no interest in running Nationals, they didn't have the time.

This was one of the major turning points in American road racing. "Fun" racers could play at Regionals, more serious competitors could contest the National Championship, paid professional drivers could build a career behind the wheel with their choice of SCCA, IMSA, international FIA and even USAC races where they could hire on to race.

Essentially, it's the same system we still have today. It has lasted so long in the same basic configuration because despite constant adjustments by the sanctioning organizations, the system *works*. Every level of racing has become increasingly competitive as the decades have gone by, but the three-tier arrangement works because it effectly separates competitors and, for the most part, pits amateurs racing against amateurs and pros against pros.

Trickle-Down

Professional road racing had a "trickle-down" effect. Last year's Can-

(Continued on page 114)

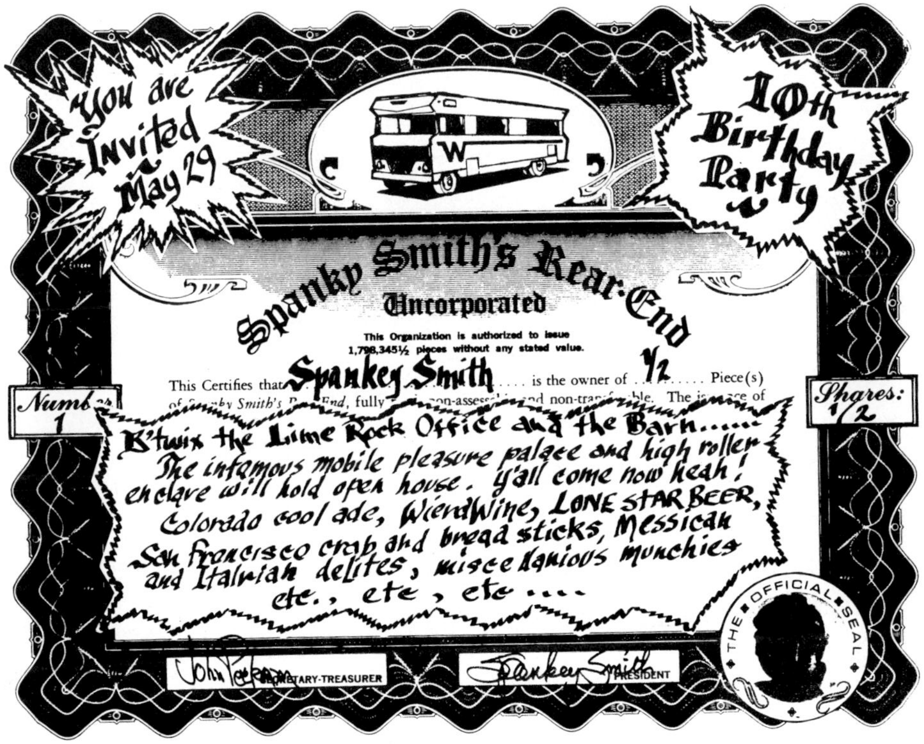

(Above) Spanky Smith, the ultimate race fan, collects an autograph. Please note the Spankster's rare belt buckle from the Polish Racing Drivers of America.
(Left) Even rarer, one-half share in Spanky Smith's Rear-End club and invitation to his Memorial Day post-race party in the Lime Rock infield behind his Winnebago. You can be sure everyone was there.

"PROMOTIONAL" RACERS

(Above) Always promotion minded, in 1967 Bob Sharp already had his transporter painted as a rolling billboard. We take such sponsor trucks for granted today, but at the time this was revolutionary in the extreme. Not even professional racing teams had such a rig, let alone amateurs.

The highpoint of "amateur" racing was reached in the mid-Seventies, when Datsun and Triumph poured literally millions of dollars into SCCA Nationals so that they could advertise their cars as National Champions. The Japanese team was led by Wilton, Connecticut Datsun dealer Bob Sharp who always carried #33 on his car; the British team by Falls Church, Virginia racer Bob Tullius who was always #44. And thus was the stage set for some of the most memorable racing battles of the Seventies, "amateur" in name only.

In an interview with Bob Cutter in 1975, Bob Sharp put it beatifully. "There are professional racers," he told Cutter. "And there are amateur racers. And then there are a few racers like Tullius and me—promotional racers, I'd call us."

"Promotional racers," according to Cutter, "are people who race in a highly professional manner for particular sponsors in whatever series might benefit those sponsors at a given moment." The best at this delicate balance between racing, advertising and public relations were Bob Sharp and Bob Tullius.

Sharp and Tullius were old friends and rivals; they'd been racing against each other since the early-Sixties. But by 1972, both had grown into racing *industries*. They were among the first to have 18-wheeler transporters for their cars, mostly because they were carrying four or six cars to all the races.

In 1972, for example, Sharp had a 240Z for C-production and a 510 for B-sedan; Jim Fitzgerald had a matching 240Z and 510. The idea was that Sharp would win the Northeast Division in both classes, Fitzgerald would win the Southeast Division and they'd both go to The Runoffs. That would give them twice the opportunity to win two Championships. By 1973, they had added Paul Newman as a driver and they won four Division Championships and qualified all five cars for the ARRC. Sharp then won his fifth National Championship, in C-production.

Group 44—more properly the Quaker State/British Leyland Racing Division—ran four cars in four different classes in those same years. A typical year was 1973: Bob Tullius in the C-production TR-6, John McComb in the D-production GT-6, Brian Fuerstenau in the E-production MG-B and John Kelly in the F-production Spitfire. They had every expectation of winning the National

Championship in four separate production classes.

Sharp vs Tullius drew thousands of spectators to SCCA Nationals, crowds that rivaled the Fifties when Nationals were as high as you could reach in road racing. The competition was even better. At Lime Rock in 1973, Bob Tullius' TR-6 started on the pole thanks to a 59.6 second qualifying lap, with Sharp next to him. Tullius led until two laps from the end, when he blew a head gasket. Sharp was right behind him, passed in The Downhill on the last lap and won. The crowd went wild.

Bob Sharp

Nobody combined being nice, being professional and being competitive like Bob Sharp. What a driver! What a salesman! This is a guy who owned a Datsun dealership and convinced his competitors, the Metro Group of 56 Datsun dealers, to sponsor his race team! His argument? Of course it would be a little better for Bob Sharp Datsun, but it would also be good for all of them. What's more, he was right.

Bob Sharp started out racing in April of 1960, when he was 21, with a Sprite and then a Lotus 7. Typically, a total unknown, he was featured in a cigarette commercial on television, which earned him a ride in a dealer-supported Datsun. At Lime Rock, on Labor Day, 1964, Bob Sharp won the first SCCA National by a Japanese car (in G-production with a Datsun 1500 roadster). His first National Championship—also Datsun's first National Championship—came in 1967, in F-production, then back-to-back ARRC triumphs in his B-sedan 510 in 1971 and '72.

Sharp opened his dealership only in 1969, purpose-built with a huge race shop out the back. Even in the early-Seventies, there were seven full-time employees preparing race cars. This was not your backyard operation, but an extremely professional racing operation that just happened to compete in events that did not pay prize money.

By 1976 when he broke his wrist in an accident at Lime Rock and retired from competition, Bob Sharp had won 45 out of 71 starts at Lime Rock,

(Left) Speeding past the old timing tower at Lime Rock, Bob Tullius in the Group 44 TR-6 in 1976. He needs a spoiler to keep the nose down at speed.

(Above) Bob Sharp inside his Datsun 510 on the grid before the Lime Rock Trans-Am in May, 1972. (Right) Leading arch-rival Brad Frisselle out of Turn Five during the Lime Rock Camel GT, Memorial Day, 1976.

22 out of 28 since 1970. Somebody figured out that he had raced more than 11,000 miles at Lime Rock alone.

Sharp segued surprisingly easily from driver to team owner. Elliot Forbes-Robinson, Bruce MacInnes and Sam Posey drove the red, white and blue Datsuns in SCCA and IMSA until Sharp settled on his friend and client Paul Newman as the perfect blend of skilled driver and promotable personality. Newman-Sharp lasted as a team at the top for more than a decade, a stability that's very rare in motor racing.

Eventually, Bob's son Scott joined Newman in the SCCA cars, before moving on to the Duracell team and the 1991 Trans-Am Championship. In the meantime, Bob had sold his dealership, closed down the race team and "retired." Which means he is busier than he ever was when he was racing.

Bob Tullius

Like his rival Bob Sharp, Bob Tullius started racing in the early-Sixties—his first race at Lime Rock was in 1962 when he entered his brand new TR-4 and led for 23 out of 25 laps before being passed by Arch McNeil's Morgan. Tullius went on to capture the SCCA National Championship in D-production in 1962, '63 and '64 with this amazing Triumph.

In 1965, Tullius quit his job as a salesman for Kodak, and he and engineer Brian Fuerstenau started Group

44. As Tullius said years later, "We approached other businesses offering our racing, marketing and public relations services as a way of promoting their products. It worked as a mutually beneficial arrangement. At the time, this was a new concept in racing.

"We were ever-mindful of our image. We never went anywhere that the cars weren't impeccable. We were one of the very first racing teams to have a full-time public relations person. I think we helped create the age of serious motorsports marketing."

By 1966, Group 44 had one of the first 18-wheeler transporters in road racing to service their four-car team. Along with Bob's TR-4 and Brian's Spitfire, they also fielded a Dodge Dart in the first year of the Trans-Am and a Lotus Cortina for Tony Adamowicz. Bob even raced the Howmet Turbine car a couple of times and had one of the most memorable drives at Lime Rock when he worked Herb Adams' old '64 Tempest into second in the rainy 1971 Lime Rock Trans-Am before blowing a head gasket.

By the early-Seventies, Group 44

had expanded into a business employing two dozen people to field a fleet of matching race cars, all sponsored by Quaker State and British Leyland. They simply dominated SCCA smallbore production racing, covering everything from C to F-production. Over the years, Group 44 cars won 14 SCCA National Championships.

In 1974, Group 44 moved up from MGs and Triumphs to a B-production Jaguar XKE V-12. This was the first factory-supported Jaguar since the LeMans efforts of the mid-Fifties. That year Tullius won the Northeast Division in B-production, sat on the pole for the run-offs at Road Atlanta, broke the lap record and finished *second* for the National Championship. In 1975, however, he came back to win The Runoffs with this car.

In 1976, Group 44 built a racing Jaguar XJ-S for Tullius and a TR-7 for John McComb. The XJ-S won both the Drivers and Manufacturers titles in the Trans-Am in 1977 and '78. This car evolved into a tube-frame XJ-S for both Trans-Am and IMSA GTO.

In 1982, Group 44 switched to IMSA GTP. In 1983, with the new Jaguar XJR-5, they were second in the championship to Al Holbert's Porsche. They raced at LeMans in 1984 and '85, and ran evolving Jaguar GTP cars in IMSA until 1987. Then Jaguar chose to bring over Tom Walkinshaw's Group C team to compete in IMSA, and Bob Tullius was without his British connection for the first time in twenty-five years. It had been one of the longest and most fruitful partnerships in motorsports.

Audi immediately picked Group 44 to run their four-wheel drive sedans in Trans-Am, winning five races in a row. That earned them a 200 lb. weight penalty and narrower tires before the Lime Rock Trans-Am, where they did not win. The Audis were unique in that they were not tube-frame cars, but built from actual production four-door sedans. It was typical of Group 44 that they were able to take such unpromising material and turn it into a race car so successful the only way to beat it was to change the rules.

For nearly thirty years, Bob Tullius and his distinctive white with green Group 44 cars have exemplified what professionalism and sportsmanship are all about...the very definition of a "promotional racer" and one of the most exciting American drivers. ■

(Above) Bob Tullius outside one of the first 18-wheelers seen in road racing paddocks, his famous Group 44/ Quaker State Racing Division transporter, at the Lime Rock Nationals in August, 1969.
(Left) Bob in the B-production Jaguar XJ-S at a Lime Rock National in September, 1976. He won both his class and overall. He's about to lap Jim Briody's A-sedan Camaro.
(Overleaf) Sharp and Tullius racing down the Lime Rock Straight on July 6, 1974. Sharp won this round of their C-production battles.

Canon
KENDALL GT-1
CP
33

44
QUAKER
STATE
TRIUMPH

Between Newt Davis, Fred Stevenson and the rest of the crew, Lotus East/Dutchess Auto was just about an extension of Lime Rock Park in the Sixties. Not to mention, Lotus Elans like this one ran away with the small-bore production races. The small, light, good-handling Lotuses could have been designed for Lime Rock's twists and turns.

Am car became this year's A-sports/racer for Nationals and eventually somebody's Regional racer. Drivers, too, would climb up the amateur ladder, switch to professional racing, then—like Peter Revson or Mark Donohue—either become international stars or more likely, run out of money or talent and slip back into the amateur ranks.

As time went on, this resulted in a tremendous blurring between "amateur" and "professional." Partly, that was because of John Bishop. Compared to the SCCA, obtaining a racing license from IMSA was almost too easy. Hundreds of inexperienced drivers were now racing seriously fast cars against paid professionals.

And IMSA paid start money to everybody. Did that make them professionals? The IRS and the racers themselves thought so. Sometimes they even convinced their wives and girlfriends they were professional racing drivers and there really was a purpose to all this expensive madness.

Then there were events like the *Car and Driver* Challenge, Rabbit/Bilstein Cup or Renault Challenge. These Showroom Stock races paid real money to amateur drivers for racing cars that cost comparatively nothing to buy and prepare. Read the driver biographies in Lime Rock race programs from the mid-Seventies, for example, and more than one VW Rabbit driver considered himself a full-time professional racing driver! No doubt some of those enthusiasts were actually eking out a living racing Showroom Stock, just one shunt away from having to go back to their day jobs.

Fast Guys and Rich Guys

It always comes down to money. The hundreds of Regional racers, then and now, were mostly paying their own bills. But there was another group of amateur racers who were not paid to drive, but who had comparatively substantial sponsorship from local or national firms.

For example, Lime Rock Expert Fred Stevenson carved himself a tremendous career in the Northeast racing Lotuses. But Fred was also a salesman for Lotus East. Not only was he racing cars that belonged to his company, every time he won a race or broke another track record, he was helping to publicize Lotus, provide more fodder for Lotus East advertisements and hopefully, sell more cars himself.

Fred ran Regionals to keep his skills sharp and show the cars to potential buyers, ran Nationals for points to get him to The Runoffs so that Lotus could advertise "National Champion" and ran professional Formula Ford and Formula B races not just for the money, but for the exposure and publicity, too.

In Fred Stevenson's case, all this was accomplished on a tight budget that limited him to racing primarily at Lime Rock and other nearby tracks. Was Fred Stevenson an amateur racing driver? Was he a professional? Both, neither, somewhere in between.

Fred is typical of many SCCA racers from this period—Mike Rand and Bruce MacInnes are among the hundreds of others—racers who had the skill to compete with the best professionals in the world, but lacked the time, contacts or finances and so were considered amateurs in everything but their driving ability. Formula Ford was chock full of these guys, demon drivers who figured they were just one lucky break away from being Formula I World Champion.

(Continued on page 119)

(Far left) The Hook lures another victim, 1975.
(Left) The Downhill tire wall has its attractions, too.
(Center) Harold Vance in an Elva Courier doo-se-doos with an Alfa Giulietta, 1964.
(Bottom far left) Another BMC engine discovers three main bearings aren't enough.
(Bottom left) This right-hand drive Honda S800 is on his second time around, but the Sprite has conveniently avoided him.

1971

May 8: Schaefer Trans-Am. The only factory car was the Penske Javelin of Mark Donohue. The best of the independents were Peter Revson and Tony Adamowicz in year-old Javelins owned by Roy Woods, Parnelli Jones and Peter Gregg in Bud Moore Mustangs, Marshall Robbins in a Camaro, Bob Esseks in Sam Posey's old Challenger and teammates Jerry Thompson-Tony DeLorenzo in Mustangs. And, of course, Bob Tullius in Herb Adams' old '64 Pontiac Tempest. Some 20,000 spectators sat in a driving rainstorm while Mark drove 200 miles in the lead to win. The story of the race was Tullius, who came from 31st—the engine had quit in practice and he didn't qualify—to second behind Donohue. Then a blown head gasket put him out 10 laps from the end. It was clear, however, that Bob Tullius could outdrive nearly anybody in the rain, and also that Herb Adams was a brilliant suspension engineer. They may have lost the race, but they both made reputations that paid off big later on.

The Trans-Am Two-Five Challenge—run as a separate race—was as competitive as the big car race. Among the leaders in a driving rainstorm were Lime Rock track owner Harry Theodoracopulos and his teammate Horst Kwech, Gaston Andrey and Bert Everett, all in Alfas; Don Pike and Russ Norborn in BMW 2002s and Bob Sharp in his Datsun 510. Almost everyone had crazy problems, from a sticking brake on Sharp's Datsun to a spin into the sandbank for Bert Everett. Horst Kwech started from the pits instead of the pole—his door wouldn't close—but by the end, he'd splashed by everyone else and lapped the field, besides. It was a day when you could do things like that.

May 29: SCCA New England Region National. One of the classic Lime Rock tales of all time; in the open-wheel race, Tom deLoughry spun in the downhill and ended up upside down in the Salmon Kill. Fortunately, he was able to unbuckle himself and swim to shore before the safety crews arrived. At the same time, Fred Stevenson was blasting through the pack trying to catch leader Mike "Pigpen" Rand. He didn't quite make it. Stevenson did win the Formula Ford race ahead of Bruce MacInnes, however, while Tom Davey won Super Vee, Bob Akroyd set a new H-production record of 1:07.3 in his Sprite and Bob Sharp won the sedan race with his 510. Sharp also won the big-bore production race with his C-production 240Z, starting and finishing ahead of all the Corvettes and Cobras. Amazing.

July 5: SCCA Northern New Jersey Region English Leather National.

September 6: L&M Grand Prix. Formula 5000 was now officially Big Time, thanks to Liggett & Meyers and a $305,000 season-long purse. Among the stars were Skip Barber, Brett Lunger, John Cannon, David Hobbs, Sam Posey, Franz Weis, Bill Brack, Eppie Wietzes and Mike Brockman. The supporting race at most tracks, including Lime Rock, was a brand-new race series called Super Vee. These were 150 mph single-seaters with 1600cc VW engines of 125 hp. Among the drivers were young hot-shoes like Tom Davey, Tom Reddy and Bill Scott.

PAUL NEWMAN

(Upper left) Happy PLN after winning the Lime Rock Trans-Am in 1986. (Lower left) Not so happy PLN beseiged by camera-wielding fans before a Lime Rock race. With only a few boorish exceptions, most racing people over the past twenty years have accepted Paul Newman as one of the gang because he is so obviously knowledgeable, enthusiastic and skillful. Plus, he's a nice guy.

Once in a while, you still hear someone say P. L. Newman isn't a "real" racer. Balderdash. At age 65, the man is still as good as any driver in the country, even winning the GT1 National at Lime Rock in 1991. Neither celebrity nor world-famous eyes will earn you multiple lap records, multiple National Championships, multiple Trans-Am wins or second at LeMans. There are full-time drivers who would run over their mothers to have Newman's racing record. As his former employer Bob Sharp says, "Paul is a phenomenon. He's still got the eyesight and reflexes of a twenty-year-old, plus tremendous powers of concentration."

Paul Newman started racing one of Bob Sharp's B-sedan 510s in 1972; he ran four Regionals with two firsts and two seconds. In 1973, he got his National license. By 1976, he was at the ARRC with an ex-Bob Tullius TR-6 with which he won the National Championship in C-production. In 1978, he replaced Bruce MacInnes in Bob Sharp's 280ZX and 200SX. He won eight out of thirteen Nationals, plus second and third at Atlanta.

The next year he was second at LeMans driving with Dick Barbour and Rolf Stommelen, then second again at the Watkins Glen 6-hour. He then proceeded to win 14 out of 16 Nationals, set a bunch of lap records and then win his second National Championship.

In 1980, he turned pro. Bob Sharp's crew chief Gene Crowe built him a 280ZX Turbo for IMSA. Newman won the first Trans-Am he entered, at Brainerd in 1983; throughout the Eighties, the team had a series of IMSA and Trans-Am Datsuns with which Newman set repeatedly lower lap records at Lime Rock. By 1985, Newman had run 53.69 to win the Trans-Am pole; in 1986 he won the race itself. At the same time, he was still running Nationals in GT1, winning his fourth National Championship in 1986. What more can you say? The man is a phenomenon. And he has credentials...real racing credentials. ■

1972

May 6: Schaefer Trans-Am. The last factory-supported Trans-Am cars were the Javelins of Penske Racing that won 8 of 10 races in 1971. With the factories out, the favorites in 1972 were the dealer-supported Javelins of Roy Woods, with lead driver George Follmer. Warren Tope, who was A-Sedan National Champion in 1971, had the best of the Mustangs, Warren Agor and Mo Carter the best of the independent Camaros. By adjusting car weights, the Two-Five Challenge was made more competitive this year. The class of the field was the BRE team of Pete Brock, that had won the Championship in 1971 with their immaculate Datsun 510s. Drivers were John Morton and Mike Downs in 1972. Their main competitors were Horst Kwech and Lime Rock's Harry Theodoracopulos in Alfas sponsored by Long Island hamburger king Herb Wetanson through his Wetson's chain. Other competitive names were Lime Rock Experts George Alderman and Bob Sharp in Datsun 510s.

May 29: Camel GT Challenge. John Bishop had run the SCCA for years until he had a falling out in 1968 and started his own road race sanctioning body, the International Motor Sport Association, racing to FIA Grand Touring rules. Lime Rock held one of the first IMSA races—a 200 miler—with ARDC Midgets as the supporting event. As you can see, the association between IMSA and Camel dates back twenty years, one of the longest partnerships in racing. So many famous names joined IMSA at the very beginning—Peter Gregg and Hurley Haywood, Dave Heinz, Mike Keyser, Charlie Kemp, Freddy Van Beuren, Bruce Jennings, Hector Rebaque, Dave White, Carson Baird. It seems like only yesterday. The support race was even more of a time-warp; the good old ARDC midgets. Among the stars were Russ Klar—who'd been in the 1959 Formula Libre race—Dutch Schaefer, Johnny Coy and legendary Len Duncan, the 62-year-old "dean" of midget racers.

July 4: SCCA Northern New Jersey Region National and Pro-Formula B Continental Championship. In addition to Bob Sharp, Bob Tullius and other drivers who were already stars in 1972, there were some entrants you might not have noticed back then but whose names you'll know now. Among them were Reeves Calloway in an Autodynamics Formula Vee, Bill Alsup in an Autodynamics Formula Ford, Evan Collins in a Winklemann, Tim Coconis in a Lotus, John Scheiffelin in a Cicada B-sports/racer, Hans Rocke in a Bobsy C-sports/racer, soon-to-be National Champion Bill Holbrook in a Royale C-sports/racer, Nick Craw in a Brabham Formula B, Frank DelVecchio in a March, Jack Cowell in a Winkelmann, Ken Caselnova in an Alfa B-sedan, Miles Collier in a Porsche E-production, Al Holbert in a Porsche C-production and Bob Speakman in a Datsun C-production.

September 4: L&M Continental 5000. John Gunn took the pole with a record 49.693; Brett Lunger won.

October 14: Car and Driver Showroom Stock Sedan Challenge. Bruce Cargill won ahead of Bruce Nole, Jack Clark and Racer Walsh.

They were so good, and it looked so easy.

There was another group of drivers with a very different problem. These were rich guys with more money that talent, who ended up racing at the highest levels of motorsports—particularly in IMSA endurance races—because they could afford to buy a ride when faster, but impecunious drivers were home working to make the next mortgage payment.

Wealthy enthusiasts who buy a ride were nothing new in racing back then, nor are they anything unusual now. Indeed, in any starting field in any level of motorsports—including Formula I and CART—a third to two-thirds of the drivers are paying for their racing, even if they are "professionals." Years ago, Sam Moses wrote a book about his own racing experiences called *Fast Guys, Rich Guys and Idiots*. That's not quite right. *Fast Guys, Rich Guys and Guys Who Would Be Rich If They Didn't Spend All Their Money on Race Cars* would have been more accurate.

Showroom Stock Sedan

For a variety of factors, amateur racing grids got thinner and thinner in the early-Seventies. It was really Showroom Stock Sedan that revitalized amateur racing. In 1972, SSS as it was called then, was strictly a Regional class brought out by the SCCA due to popular demand from club members who were looking for a way to go racing on the cheap in safe, reliable cars.

For the first four years, SSS was a true entry-level class that almost anybody could afford. For the first time since 1959 or so, it really was possible to drive to the track, stack your luggage in the paddock, race to win, put your luggage back in the trunk and drive home. The only drawback to SSS, like most other race series, was that you had to pick the right car to be competitive. And the right car changed every year.

All too soon, Showroom Stock cars started rolling out of eighteen-wheelers, and there were batteries of full-time mechanics in the pits checking tire pressures after every session (there wasn't much else to check on a Showroom Stock racer). The serious Showroom Stockers were soon moving suspension pickup points for better handling, experimenting with tire compounds and applying every trick

(Continued on page 122)

(Below) Mike Rand heads off for another victory lap in Formula C in 1972. Behind him is the old timing and scoring tower on the outside of the Main Straight.

What was so neat about the Car and Driver *Showroom Stock Sedan Challenge? For one thing, it was professionally produced and promoted down to the smallest detail—Art Director Gene Butera even designed handsome IDs like this corner worker button for everyone involved—there was free beer for the spectators, which gives everyone's memories a roseate glow, the "readers" versus "writers" rivalry added an extra dimension to the contest and most of all, watching little cars slowly bang into each other and roll over and stuff was harmless* fun.

CAR AND DRIVER CHALLENGE

The way I remember it, and I was *Car and Driver* Managing Editor at the time, it was all Pat Bedard's fault. The SCCA had proposed a Regional racing series for Showroom Stock sedans in 1972, and Pat saw an opportunity to 1) Go racing, 2) Publicize the magazine, 3) Generate some articles, 4) Do something with the totally-trashed long-term Pinto and Vega test cars we were embarassed to give back to the manufacturers and 5) Throw a hell of a party on Bill Ziff's expense account. The whole thing was to be sort of an end-of-season ARRC for cars that weren't allowed to go to The Runoffs, only better.

It quickly got out of hand. Bedard and his assistant, Don Sherman, spent all spring and summer testing little sedans, trying to find The Secret Weapon. That turned out to be an Opel, so that's what Bedard got himself. Don Sherman had to make do with a Pinto that already had 30,000 miles on it running errands in mid-Manhattan. Jim Williams talked a Colt out of Brock Yates' driveway only a week before the race. At the time, I was road racing Yamahas with the AMA, and considered competing in something as declasse as a Pinto that ran 1:14s at Lime Rock to be downright laughable. And laugh we did.

As Bedard wrote at the time, "When the green dropped, cars fanned out as if they had been sprayed from a hose. They were everywhere, weaving and sliding, their fenders bouncing together." It was The Three Stooges Meet Bumper Cars. There were heats to reduce the 77 cars that showed up—from as far away as California—to a more manageable thirtysomething for the final. Bedard and Sherman were both involved in accidents and taken out; Williams motored along to finish sixth. Bruce Cargill won $3300 and the applause of the crowd of 7500, now thoroughly lit up on free Schaefer beer. Bruce Nole, Jack Clark, Jerry "Racer" Walsh and Gary Grochow were the rest of the Top Five.

Absolutely the highpoint of the whole weekend was Saturday night when expatriate British photographer Humphrey Sutton, who'd been hanging around the bar all day between races, stood up on the table at Stagecoach Hill Inn, raised his flagon high and yelled "Right you are if you think you are" before passing out on the table in a clatter of flying glasses and broken dishes. It was that kind of weekend.

The next morning, a vengeful Sutton took the infamous "Deadly Photograph" of the *Car and Driver* Staff that was our corporate Christmas card in 1972. By 1976, only three of the eighteen people in the picture were still at the magazine. Incredibly, those same three—Brock Yates, Pat Bedard and Marketing Services Manager Bill Strock—are still there.

I was gone before Challenge II, now a *C/D* Contributing Editor so that I could get paid real money to do what I had been doing before for free racing hats and a handshake. Pat Bedard had become a real racing driver by getting Bill Ziff to fund a Mazda for IMSA that was so fast it won three poles, two races and was promptly banned, we had all traded in our C/D denim racing jackets for red-white-and blue vinyl racing outfits that were so hot we called them the Linoleum Jackets and Showroom Stock had exploded into the most successful racing series in history. Bedard got his revenge by winning his heat and also the final in front of Paul MacDonald, Edwin Allen, Paul Hacker and Zane Lumbley after Don Knowles' Opel broke. At Challenge III, Bedard won again after a race-long battle with Knowles, with Joel Bradley third and Don Sherman fourth ahead of Steve Behr.

Challenge IV in 1975 was the year it rained. And rained. Among the big names who ended up back in the pack were Al Holbert, Bruce Cargill, Sam Posey and Patrick Bedard. Paul Hacker beat Don Knowles, Patrick Jacquemart and Don Sherman. The *C/D* staff spent most of our time stifling in Linoleum Jackets in the VIP tent next to the hot coffee urns. It was not a vintage year.

By the final *C/D* Challenge in 1976, Showroom Stock had escalated. Saab imported Swedish rally star Stig Blomqvist to drive a 99GL. Blomqvist lowered the SSB lap record to 1:12.4 and beat Bedard in an identical car, with Joel Bradley, Sherman, Hacker and Knowles behind them. Don Sherman finished fourth in three Challenge races which—aside from giving Showroom Stock legitimacy at a critical time—might be *C/D*'s most lasting contribution to racing. ■

(Top left) Challenge IV winner Paul Hacker leads Don Knowles, Patrick Jacquemart and Don Sherman in the rain, October, 1975.
(Left) Challenge I, October, 1972. Remember when an Opel Manta was the hot set-up? This one's just gotten bit by The Hook while chasing a Pinto. A Pinto?
(Above) Two drivers who got their start at the C/D Challenge. Pat Bedard (left) rose as far as the Indy 500; Don Knowles dominated SCCA Showroom Stock.

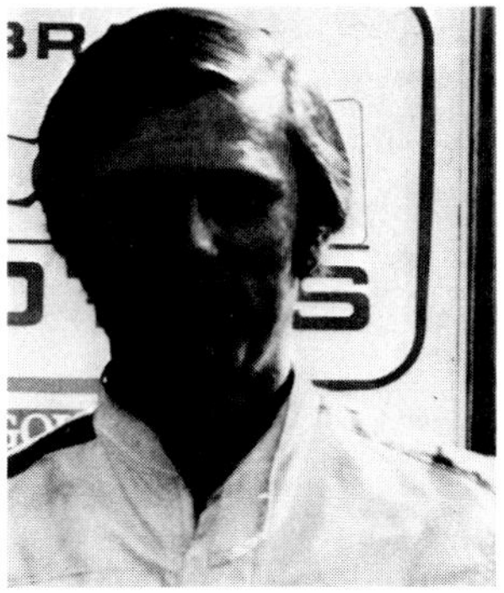

in the racers' handbook to gain an advantage. For some people, that was the fun of it, trying to outsmart the competition and the tech inspectors. Other racers saw constant development as a problem.

The Scirocco/Bilstein Cup attempted to solve that problem. Jo Hoppen, competition manager at Volkswagen of America, who had already helped make Formula Vee and Formula Super Vee so successful, came up with the idea of racing identical Sciroccos in a series sponsored by the Bilstein Shock Absorbers.

It was a brilliant plan. Since the cars were as identical as could be, who went faster was up to the driver...and the mechanic who prepped the car, of course. Unlike most road racing, Scirocco/Bilstein Cup races—and the subsequent Rabbit/Bilstein Cup races—were fun for both the enthusiastic competitors and spectators alike.

If the Bilstein Cup was fun, the Renault Challenge was hilarious. This was one-design racing stretched to an absurd degree, with little Renault LeCars tip-toeing around Lime Rock at impossible speeds. Mostly, racing LeCars roared three abreast into The Hook or bump-drafted down the Straight. It was the most entertaining race series there could possibly be, and proof once more that what makes a good specator event is not speed, but close competition. Thank Renault's engaging competition director, the late Patrick Jacquemart, for making it all so enjoyable.

Lime Rock Amateurs

Lime Rock turned out to be the center of amateur road racing not just for the Northeast, but really for the whole country. Perhaps that was because, from the very beginning, it was run by and for amateur racers. John Fitch may have been a famous member of the Mercedes factory team, but at heart he was a gentleman amateur racing driver who competed because he loved it. Track manager Jim Haynes was even more of a die-hard amateur, as was track owner Harry Theodoracopulos. They raced for the fun of it, and they made it fun for other amateurs to go racing at Lime Rock, too.

Jim Haynes, in particular, loves

(Continued on page 127)

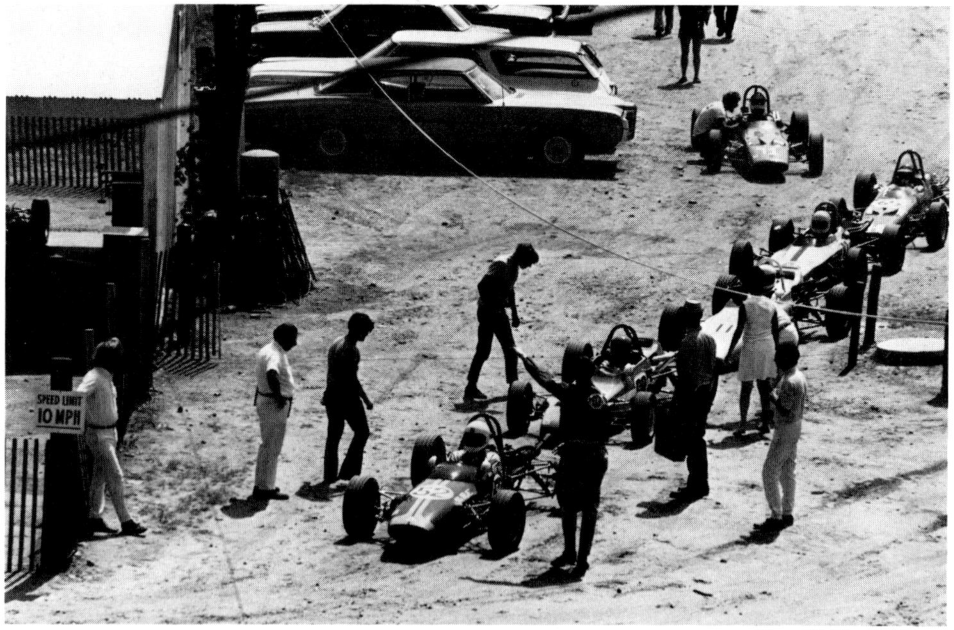

(Above) W.P. "Fred" Stevenson was probably the most expert of all Lime Rock Experts, holding class lap records at least eleven different times over the years. (Right) A Formula Ford grid heading out for qualifying in 1972.

FORMULA VEE

The first two were Nardi Formula Juniors built in 1959 and fitted with 1100cc VW engines for a Jacksonville, Florida Regional racer/VW dealer named Hubert Brundage. Pretty soon, George Smith and Bill Duckworth were building cars they called "Formula Vee." The SCCA recognized the class in 1963: 1200cc VW single-port engines and stock transaxles, stock VW front suspension, brakes and wheels, all put together with a simple tubular frame and a fiberglass single-seat body.

Unlike Juniors, which became the "car of the month club," Vees evolved very slowly. A Vee was the perfect starter car for beginning racers, it was the perfect experimental class for demon homebuilts, it was the perfect vehicle for learning how to keep your foot in it and never lift. Other racers made fun of Vees as "angry overshoes," but at a tight track like Lime Rock the best Vees could circulate as quickly as D-production sports cars that cost far more to buy, race and repair. Even today, as Vees celebrate their 30th Anniversary, they are still the best bargain in racing. ■

A mixed gaggle of Formula Vees heads onto No Name Straight during the July 6, 1974 SCCA National.

1973

May 5: Schaefer Trans-Am and VW Gold Cup. Even the Lime Rock program, *Racing* magazine, labelled the revised Trans-Am "Car(icature) Racing." In response to declining sponsor support and the successful IMSA Camel GT, the SCCA switched to FIA Group 2 and 4 rules just like IMSA. In essence, this meant up to 460 cubic inch big-blocks in the Trans-Am sedans, plus the addition of big-block Corvettes and 2.5-liter Porsche 911RSs. The Secret Weapon turned out to be the 2.9-liter German Capri V-6 that dominated the European Touring Car Championship for years. Among the Trans-Am leaders were Peter Gregg and Al Holbert in Porsches, Marshall Robbins in a Corvette, John Buffum in his famous Ford Escort, Gene Felton and Charlie Kemp with Camaros. The support race was Super Vee; Tom Davey and Tom Reddy were already the guys to beat. As of May, 1973, the lap records at Lime Rock Park were:

AP: Kepler, Corvette, 1:00.4
BP: Anderson, Corvette, 1:01.3
CP: Sharp, Datsun, 1:01.0
DP: Wagner, Elva, 1:03.4
EP: Kraftson, MG-B, 1:03.8
FP: Kelly, Spitfire, 1:04.0
GP: Hauser, Sprite, 1:06.1
HP: Canfield, Sprite, 1:06.7
ASedan: Posey, Mustang, 59.2
BSedan: Sharp, Datsun, 1:01.2
CSedan: Spreen, Mini-Cooper, 1:05.4
DSedan: Gross, Mini-Cooper, 1:07.4

FA and track record: McRae, McRae, 50.0
FB: Sarich, March, 52.2
FC: Mitchell, Brabham, 57.9
FV: Laughlin, Stohlman/Zink, 1:04.0
FF: MacInnes, Winklemann, 57.4
FSV: Reddy, Lola, 55.6
ASR: dePasquale, Lola, 56.4
BSR: Mitchell, Lola, 55.2
CSR: Consentino, Fiat-Abarth, 59.1
DSR: Igleheart, Bobsy, 105.4

July 7: SCCA Northern New Jersey Region National. Among the more recent converts to racing at Lime Rock; Patrick Bedard, Bob Koveleski, Ken Heath, David Ammen, Norm Marx, Janet Guthrie and Paul Newman.

September 15: IMSA Camel GT. Winner of the Radial Challenge race was Pat Bedard in the *Car and Driver* Mazda RX-2, the first Mazda and the first rotary to win a professional race in America. Bedard also set the lap record at 1:09.2. IMSA promptly changed its rules to make the Mazda uncompetitive. Bedard came back the next year in a Pinto. *Sic Transit Gloria.*

October 13: Car and Driver Challenge II. Pat Bedard won with an Opel.

BRUCE MACINNES

"Sometimes, talent isn't enough." That headline has been part of the baggage Bruce MacInnes has carried through life since 1985, when *AutoWeek* put him on their cover. But the reality goes back twenty years.

What did *AutoWeek* mean? Only that Bruce MacInnes is arguably the most skillful driver ever to sit behind the wheel of a racing car. And yet he never got that phone call from Enzo Ferrari to meet him at the bar of the Hotel de Paris in Monte Carlo to talk about his future as Formula One World Champion.

Why? Naivety. No financial angel. Not the right connections. The wrong series. Who knows? When MacInnes started, Formula Ford was said to be the road to the top. He was IMSA Pro-Ford Champion in 1972, AFFA Pro-Ford Champion in 1976. He led the SCCA Northeast Division in 1972 and '75-'77, and started from the pole at The Runoffs three years in a row. He held 18 lap records and most seasons won every race he entered.

In 1977, Bob Sharp put Bruce in both his C-production 280Z and his B-sedan 200SX. MacInnes won eight of nine starts, but both cars broke at The Runoffs. Says Sharp, "Bruce is a great driver, a *very* talented guy." Sometimes, talent isn't enough.

Somebody read that nice *AutoWeek* piece and put Bruce in an out-classed Can-Am car at Lime Rock Park. Bruce won the 1985 Lime Rock Can-Am in a rain storm so bad they stopped the race in the middle. Since then he's won the Nelson Ledges 24-Hour race in a Consulier and done some Supercar races. But nothing that's going to get him to Monaco in a Ferrari.

Bruce is very goal-oriented; he wanted to race Formula I and he wasn't going to get there racing Nationals. So he set himself a new goal and joined Skip Barber's organization in 1977. He set out to be the best racing instructor in the world. His students say he's accomplished that; Skip Barber School Instructor Bruce MacInnes is acknowledged one of the best teachers of racing to be found.

As Bruce asks his Skip Barber students, "Who was World Champion in 1970?" Nobody remembers (it was Jochen Rindt). It's taken a while, but Bruce MacInnes has learned an important cosmic secret. Life is not what F.M. Alexander calls *end-gaining*; life is gaining the *means whereby*. Or as William Least Heat Moon puts it in *Blue Highways*, "Any traveler who misses the journey misses about all he's going to get." ■

(Above) Bruce after winning the Lime Rock Can-Am in 1985. Seymour Chicken waited in the Motorhome.

1974

May 4: Schaefer Trans-Am and Gold Cup Super Vee, Vintage Sports Cars, The Great Schaefer Pedal Race plus Sky-Diving. By now, the Trans-Am and IMSA GT had become essentially interchangeable, with stars like Harry Theodoracopulos with his German Capri, Warren Agor with his orange Camaro, Al Holbert with his RSR Porsche, John Buffum with his BMW, Warren Tope with his Pantera, Sam Posey with Ted Trudon's Porsche Carrera, Peter Gregg with his Brumos Porsche RSR, Jocko Maggiacomo with his Javelin, Elliott Forbes-Robinson with his yellow Porsche Carrera, Ludwig Heimrath with a Porsche 911 and Paul Newman in Tom Ciccone's British Ford Escort. Al Holbert won the Lime Rock Trans-Am in his Porsche, after Peter Gregg and Michael Keyser crashed on the first lap. Elliot Forbes-Robinson won Super Vee.

The VSCCA Vintage Race included such well-known racer/collectors as Bob Herlin (MG-K3), John Gelles (OSCA), Jim Haynes (Frazer Nash), Bob Akin (Lotus 11), Joel Finn (Maserati Tipo 61), Tony Koshland (OSCA), Judy Stropus (OSCA), Steve Wilder (Porsche 550), Bill Gelles (Ferrari 250 MM), Homer Tsakis (Siata 1100), Gaston Andrey (Maserati Tipo 61) and J.D. Igleheart (OSCA).

July 6: SCCA Northern New Jersey Region Kendall Cup National. Among the interesting entries were future IndyCar aerodynamicist Tony Cicale in a B-sports/racing Chevron, Peter Symonds driving Stutz Plaisted's twenty-year-old Cadillac-powered Cheetah and a C-production field that included Bob Speakman, Beck Beckwith, Alby Kalajzich and Bob Sharp in Datsuns versus Bob Tullius and Dave Manners in TR-6s. Among the defending Northeast Division Champions were Sam Feinstein (A-production Cobra), Bob Sharp (C-production 260Z), Brian Fuerstenau (E-production MG-B), John Kelly (F-production Spitfire), Mike Dale (H-production Sprite), Eric Kerman (C-sports/racing Arachnid) and Ken Duclos (Formula B Brabham). There was a big brouhaha over the Formula Ford race; 60 racers showed up for 33 starting spots. Half were forced to qualify in the rain, and were sent home.

September 2: IMSA Camel GT Challenge, Goodrich Radial Challenge, Gold Cup Super Vee. Porsche won 10 out of 12 GT races in 1974, which was mostly the reason that John Bishop invented All-American GT to give US-based cars a break. Porsche retaliated with the wild 935, of course, preserving the status quo but making for some incredible races. Gregg won the first heat of the Camel GT with Michael Keyser second and Steve "Yogi" Behr in third; Keyser won the second heat ahead of Bob Sharp and Milt Minter.

Among the Super Vee contenders were Tom Reddy, Howdy Holmes, Reeves Callaway, Tom Bagley, Elliot Forbes-Robinson, Hurley Haywood, Bill Alsup and Bertil Roos. Harry Ingle beat Bill Scott, Fred Phillips and Rickie Bell. In the Goodrich RS race, Nick Craw in a BMW 2002 beat Amos Johnson and George Alderman in Gremlins and Jerry "Racer" Walsh in a Pinto.

October 10: Car and Driver Challenge III. Pat Bedard wins for the second year in a row, driving a Vega.

formula cars. Under his reign, Lime Rock was always a special center of open-wheel racing. Besides, Lime Rock is especially suited for racing drivers who don't race for a living. The track is a fun spot to spend a family weekend, it is easily accessible from almost anywhere in the Northeast and the course is slow enough not to scare a neophyte yet challenging enough to keep drivers coming back year after year. It's also far and away the nicest race track in the country to which you can bring a major business client, most of whom are from nearby New York.

This nearness to Manhattan is at least part of the reason Lime Rock hosted the annual *Car and Driver* Challenges, and also why the Renault Challenge races headquartered here. Can you imagine racing LeCars at Road America or Sebring? Lime Rock's tight little corners might have been especially designed for Showroom Stock as well as small formula cars. It's always been the perfect amateur racing venue.

The Regions

Another factor that has made Lime Rock Park such a special place for amateur racing is that, unlike most race tracks, SCCA races at Lime Rock are sanctioned not by just one region, but by New England, New York, Northern New Jersey and Mohawk-Hudson regions of the SCCA. There have usually been at least two SCCA Nationals at Lime Rock each year, plus half-a-dozen Regionals. If it were up to the Regions, there'd be more.

Among other things, having four strong Regions racing here has meant that Lime Rock has always had the best corner workers, grid marshals, chief starters and stewards in the

(Left) The famous All-Girl pit crew of Boffo Motors—drivers Tom and Jim Boffo—at work preparing their matched Minis in 1972.
(Below) Sam Feinstein on his way to winning A-production at Lime Rock Park in July, 1974 in his 427 Cobra.
(Bottom) Alex Dearborn ran his dune buggy as a sports/racer and beat the McLarens! Sometimes.

country. This abundance of talent has also meant that it has been comparatively easy to find workers for non-SCCA events, just because there is such a pool of skilled people to draw from. From Bonnet & Boot to Skip Barber School, all sorts of organizations have availed themselves of racing at Lime Rock.

That has helped make Lime Rock an amateur racing center, too. The typical neophyte can start by driving around the track with the BMW Club, say, then graduate to Bonnet & Boot. That gets him interested in improving his skill level, so he attends the Skip Barber School. That might then lead to a Formula Ford or a Showroom Stock car...and presto! Another amateur racing driver is created.

Amateur Allure

If you examine the entries for Lime Rock Nationals and Regionals over a period of time, you discover that the typical amateur racing driver competes for seven years. After that, they either run out of time, run out of money or simply get burned out from the incredible concentration that racing demands. To be sure, there are drivers like Chuck Cunningham, George Alderman, J.D. Igleheart, Charlie Gibson, or yes, even Bob Grossman, who've raced *something* at Lime Rock every year since it first opened. But by and large, the thousands of drivers who've competed in SCCA events at Lime Rock over the decades have stayed in racing for a comparatively short time. The ones who keep at it year after year are very much the exceptions.

What possible allure can there be for an international star like Bob Grossman to run Lime Rock Regionals in a Showroom Stock Camaro 35 years after he first raced here? Ah, now you're getting at the heart of what it means to be a racing driver. Bob Grossman, like most other racing drivers, started because he liked cars, he liked driving fast and it was fun to compete. He also turned out to be very good at it, good enough to advance from Lime Rock to LeMans, which is a very long way, indeed. But for Bob, it was always a pastime.

While no one would argue against being paid to race, the professional driver is there for only one thing. To win. You can't come back after a race that cost your sponsor something like $2000 a mile and say, "Boy, did I have fun! And so what if I'm not in the Top Ten, at least I finished and we all had a great weekend." Neither can you say, "I don't feel like racing this weekend. Let's just stay home."

The amateur driver, on the other hand, gets all the pleasure with none of the pressure. He gets to race when he wants, how he wants, where he wants. Any pressure is self-induced...unless you've got a pushy volunteer crew chief, but that's another story. The motto of amateur racers is "We pay to do this for fun." That sometimes takes a lot more dedication than when you're getting $17-million a year to do the same thing, but the stress level is a lot lower.

And that's what somebody like Bob Grossman is doing driving in amateur races at Lime Rock. Having fun. There is nothing you can do that is more enjoyable than racing just for the fun of it...racing because you want to, not because you have to. Do well, and there is an immense feeling of satisfaction, of self-confidence, of well-being that puts a smile on your lips and a spring in your walk for days afterwards. As Connie Kalitta says in the movie *Heart Like a Wheel*, "Racing isn't *golf*, you know." Precisely.

5.
IMSA: Super Pros

(Above) Sam Posey leads Hans Stuck in their factory BMW 3.0 CSLs into West Bend during the 1975 Camel GT.
(Right) Peter Revson on the Camel GT grid at Lime Rock on September 15, 1973. He was "guest driver" in the Harvey Wallbanger Porsche Carrera.
(Previous page) Peter Gregg in a 1975 Brumos Porsche 935; IMSA personified.

The International Motor Sports Association started because John Bishop, who was the Chief Executive running the whole SCCA, got into a disagreement over the proper way to develop the Professional Racing Division. The SCCA, after all, was still a club for weekend warriors, most of whom weren't sure they even needed or wanted a Professional Racing Division to sanction Bishop's Can-Am, Trans-Am and Continental Championships. Bishop agreed with them. He believed that the SCCA should stick to club racing and that there should be a separate organization to handle professional racing.

In the May, 1969 *Car and Driver* there's a headline "Bishop Resigns in SCCA Shakeup." By then, he had already organized IMSA and was ready to start holding professional races in direct opposition to the SCCA. John Bishop is nobody's fool. Among other things, he got FIA approval for his races, so that the SCCA could not bar SCCA racers from participating in IMSA events, something they were all too likely to do. In 1972, he got Camel to come on board as his series sponsor...IMSA is still running the Camel GT two decades later.

Start Money

One of John Bishop's great innovations was to pay start money as well as purses. This meant that at the end of the day, when everyone was tired and thirsty and wondering how the hell they were going to afford to hang four new fenders *and* rebuild the top end, John Bishop would have this little ceremony and hand out checks with a smile and a handshake to every team that competed.

The checks might only be for $75 by the time you got down to nineteenth place, but as Milo Minderbinder promised in *Catch 22*, "everybody got a share." There were racers who would have killed for John Bishop, just because of those little checks they could bring home to show their wife...and their accountant. John Bishop made *everybody* a professional racing driver. Particularly at tax time, that was worth its weight in Porsche Turbos.

Bishop adopted FIA Grand Touring rules so that a team could run the same car in IMSA races and also at Sebring, the Daytona Continental, Watkins Glen or even, if they could afford it, LeMans. There was a Grand

Touring class for FIA Group 4 sports cars—divided into Over 2.5-liter cars and Under 2.5-liter cars. There was an FIA Group 2 Touring class—essentially for four-seater Trans-Am sedans—plus an Under 2.5-liter Touring class for the Porsche 911s and BMW 2002s that were running the small-bore Trans-Am. Finally, Bishop added a "Baby Grand" class that was somewhere between Showroom Stock and Under 2.5 Trans-Am.

(Left) John Bishop has style; his Camel GT pace car was a DeTomaso Pantera in 1973! Warren Tope tried to turn a similar Pantera into an IMSA GT car, with disappointing results.

Benevolent Despot

Most importantly, John Bishop realized that what was wrong with the SCCA was that they were trying to run three major professional racing series by committee. In the SCCA's case, a committee of 20,000. Can't be done. Bishop took for his model Bill France of NASCAR, whose dictatorial reign had led to stable rules, terrific racing, overflow crowds, happy sponsors and millionaire racing drivers.

If you were John Bishop in 1969, comparing the way the SCCA and NASCAR were run, there was really no decision which you'd rather be. Bishop actually got Bill France to put up 25 percent of the money to start IMSA. As *Car and Driver* put it at the time, "IMSA operates without benefit of the procedures outlined in *Robert's Rules of Parliamentary Procedures.* John Bishop is the benevolent despot of IMSA." As it turned out, a benevolent despot was just what road racing needed.

After a somewhat shaky start, by 1972 IMSA had taken on much of the upscale feel it would have throughout the next two decades. By 1973, Bishop had expanded in all directions, some of which turned out to be good, some of them not so good. One of the latter was the short-lived Century Series, also known as Pro-Formula Ford. IMSA was most successful when it stuck to racing high-performance coupes and sedans.

RS

By 1973, Bishop had gotten a sponsor for his little "Baby Grand" sedan series, which now became the B.F. Goodrich Radial Challenge for RS (Racing Stock) cars. This was one

1975

Skip Barber started his driving school and ran his first classes at Lime Rock Park.

April 26: SCCA New York Region North Atlantic Road Racing Championship Regional.

May 26: IMSA Camel GT, Goodrich Radial Challenge and Robert Bosch Gold Cup. Peter Gregg in his Porsche RSR beat Hans Stuck (BMW), Al Holbert (Porsche) and Sam Posey (BMW) in the first heat of the Camel GT; Holbert won over Gregg, Stuck and Posey in the second heat after Gregg spun on the last lap. Holbert won overall. Lime Rock Park owner Harry Theodoracopulos destroyed his Chevy Monza in practice and drove the Pace Car. Bob Sharp won the Under 2.5-liter class; Eddie Miller won Super Vee. Carson Baird beat Dennis Shaw, Amos Johnson and Nick Craw in the RS race. The most exciting moment of the day was when Baird was almost blown off the course on the next to last lap thanks to a strafing run by the Goodyear Blimp.

July 5: SCCA Northern New Jersey Region Kendall Cup National. Among the neophyte racers in the Showroom Stock race were Eric Morrow from nearby Sharon, writer Bill Lovell, John McKnight is his VW Wabbit entered by Elmer Fudd, Steve Smith and David Rosenblum. Other soon-to-be-famous names included David Ammen and John Stim in C-sedan, Ed Triolo in Formula Vee, David Loring and Norm Marx in Formula Ford, Janet Guthrie in B-sedan and Paul Choiniere in D-production. Among the stars in A-sports/racing were Oscar Koveleski with his McLaren and Eno DePasquale with his Lola.

July 26: SCCA New England Region NARRC Regional.

August 24: SCCA Northern New Jersey Region NARRC Regional.

September 1: SCCA National. As of 9/1/1975, the Lime Rock lap records were:

AP: Kepler, Corvette, 1:00.4
BP: Tullius, Jaguar, 59.6
CP: Sharp, Datsun, 59.6
DP: McComb, Triumph, 1:01.5
EP: Frey, Porsche, 1:02.1
FP: VanderVate, Spitfire, 1:02.3
GP: Hauser, Sprite, 1:03.8
HP: Canfield, Sprite, 1:04.6
ASedan: Posey, Mustang, 59.2
BSedan: Sharp, Datsun, 1:02.4
CSedan: Henderson, Datsun, 1:03.2
DSedan: Gross, Mini-Cooper, 1:07.4
FA and Track Record: McRae, McRae, 50.0
FB: Duclos, Brabham, 51.9
FC: Wrightson, Brabham, 55.2
FV: Van de Car, Caldwell, 1:02.3
FF: Bell, Crossle, 56.7
FSV: Bagley, Royale, 55.3
ASR: dePasquale, Lola, 55.0
BSR: Cicale, Chevron, 53.8
CSR: Stevenson, Bobsy, 56.0
DSR: Igleheart, Bobsy, 105.4
SSS: Wright, Rabbit, 1:13.4
SSC: Honeywell, Porsche, 1:12.8

September 13: SCCA Northern New Jersey Region NARRC Regional.

October 11: Car and Driver Challenge IV. Paul Hacker in a Scirocco evened the score to Readers 2-*C/D* 2.

of the best racing series ever invented. You could get into it for almost nothing—Joe Varde pulled his Gremlin out of a junkyard—you could run anything from a BMW 2002 to, well...a Gremlin, and you could win enough money to actually pay for your racing. Compared to the SCCA's amateur Showroom Stock series, IMSA's professional Baby Grand races were a red hot, fender-banging hoot.

Camel GT

The Camel GT races were even better. Running at the front you'd have Corvette Stingrays with all-aluminum big block 454s putting down 700 hp. And Trans-Am-style Camaros with 460 cubic inches. And Porsche 911Ss and 914/6s with 2.5-liter Sixes. Because Bishop was following the NASCAR formula, he would juggle the weights to make everybody's performance just about equal.

But the Camel GT Championship was made up of ten long *endurance* races, at least partly because endurance racing garnered better press coverage, required more drivers and attracted a higher level of participants. When it came to the races themselves, the Porsches were just as fast as the Corvettes but more durable...and they got much better gas mileage.

That reliable frugality made the Porsche racing team of Peter Gregg and Hurley Haywood national heros and probably did more to boost Porsche sales in the United States than anything except Steve McQueen driving a Gulf-Wyer 917 in the movie *LeMans*. In those days, Porsche *owned* endurance racing on both sides of the Atlantic. Their success in IMSA was a big contributor to that all-conquering reputation.

All-American GT

In 1974, John Bishop invented another terrific racing series within the Camel GT which he called All-American GT. Basically, this was a way to give the American cars a break in terms of weight, engine size and aerodynamic bodywork. It led to things like the Greenwood Corvette, DeKon Monza and Charlie Kemp's crazy Mustang II, all covered with wings, fins, flares and housing serious V-8s. All-American GT was so successful that within a year, Greenwood Corvettes and DeKon Monzas dominated IMSA.

As Porsche retaliated with the 935 Turbo, Bishop made sure that the American cars maintained their parity. It may not have always been fair—Peter Gregg threw a fit in 1975 because IMSA outlawed the rear wing on his Porsche—but like NASCAR, John Bishop's rules juggling gave every team a chance, every manufacturer something to use in their advertising and every race fan something to talk about on Monday morning. IMSA, from 1970 through the late-

(Continued on page 140)

Camel GT, Memorial Day, 1977: Two BMWs lead a Porsche Carrera through the Esses. The hill is jammed with spectators on this hot and sunny day. Notice the debris and Speedi-Dry in the corner from an earlier incident.

KENDALL
GT
GOODYEAR

Brumos
Porsche

Eighties, was the most successful professional road racing group in the world.

Peter Gregg stops for gas and tires in his BMW 3.0 CSL during the 1976 Lime Rock Camel GT.

Super Vee

IMSA has always been essentially a sedan and sports car racing group. But when Super Vee came along in the early-Seventies, IMSA, like the SCCA, embraced the class with tremendous success. The whole idea of Super Vee was to have a transition class for formula car drivers that would move them from the starter classes of Formula Vee and Formula Ford and into Formula 5000, Indy cars or even Formula I.

Super Vee had the backing of Volkswagen and Bosch, which obviously didn't hurt. Think of the cars as Formula Fords with hot-rodded VW engines, capable of 140 mph even twenty years ago. All sorts of neophyte formula car drivers advanced their careers by running in Super Vee, from Howdy Holmes and Al Holbert to Bill Alsup and Bertil Roos. For the better part of two decades, Super Vee did just what VW's Jo Hoppen intended when he formulated the class; it gave aspiring professional drivers a place to get experience, it promoted the Volkswagen name and it provided *great* racing.

Kelly American Challenge

Another one of John Bishop's brilliant ideas was his American Challenge, introduced in 1977. This was really just another variation on the old Trans-Am idea, running mid-size American V-8 sedans on road courses. In the American Challenge format, the cars were limited to 366 cubic inches in 3300 lb. cars.

The first two years, the cars were run as merely one class within the Camel GT races. But for the 1979 season, Bishop came up with Kelly Services as the series sponsor and started running separate Kelly American races. There was a special points fund for women drivers, and when Lyn St. James finished second to Gene Felton at Road Atlanta in April of 1979, it seemed like the whole "women's racing series" idea would be an absolute gold mine.

It never quite worked out that way. Gene Felton absolutely dominated Kelly American at the beginning, then drivers like Vern Smith, Craig Carter, Irv Hoerr and Tommy Riggins took over. Women drivers finished near the front, but never seemed to win. No matter. The women were competitive racing fender to fender in a top-rank series; the days of Powder Puff racing were over.

What Kelly American did was encourage stars like Lyn St. James, Patty Moise, Kathy Rude and Deborah Gregg to go racing and then advance into more competitive series like Trans-Am, IMSA GTO and in St. James' case, the Indy 500. Even if that had been Kelly American's only accomplishment, it would have been a worthwhile series.

IMSA Stars

Because it is a club primarily designed to serve its amateur members, the SCCA sometimes seems to resent stars who rise above the norm. John Bishop's IMSA had no such inhibitions. IMSA always attracted wealthy, photogenic racers like Peter Gregg, Hurley Haywood, Rob Dyson, Bob Akin, David Hobbs, Deborah Gregg, Lyn St. James, Paul Newman and Al Holbert, people who are, in *Town & Country*'s happy phrase, "socially acceptable."

These socially acceptable racers gave IMSA instant cachet. Starting from scratch, within five years IMSA was the best known, best-regarded road racing organization in the United States. There was a place for everybody, from struggling racers like Joe Varde or Amos Johnson who were trying to make a living and a reputation at a tough game, to instant stars like Gregg and Haywood who brought as much to IMSA as IMSA gave to them. If the Harvard Business School wanted to do a case study on how to start a professional racing series, all they'd have to do is look at IMSA in the Seventies.

IMSA not only brought class to American road racing, it brought a degree of professionalism that had been missing before. Cars prepared by Brumos, or Holbert's, or Bob Sharp or even Gianpiero Moretti's Momo

(Continued on page 146)

Kelly American Challenge never became the "woman's racing series" that some proponents envisioned, but it did jump-start a lot of careers for both men and women. Pat Bedard and Lyn St. James made it all the way from Kelly to the Indy 500, for example. This is the Kelly grid at Lime Rock in 1979, mostly filled with Chevrolet Novas and similar GM derivatives.

1978

May 29: IMSA Camel GT, Champion Spark Plug Challenge and VSCCA. Revised IMSA rules comprised three classes of Grand Touring—essentially the same as the previous GT class but incorporating GTU as well—Grand Touring Experimental—meaning turbocharged Porsche 935s and McLaren BMWs—and All-American GT—the familiar Monzas, but now with ultra-wide wheels, 120 lbs. less weight and up to 455 cubic inch V-8s. In addition, of course, there was the RS Challenge, with new series sponsor Champion.

Among the big names in IMSA at this time were Dick Barbour, Phil Currin, Peter Gregg, David Hobbs, Bill Whittington, Bonky Fernandez, Ludwig Heimrath, David Hobbs, John Paul, Gianpiero Moretti, Kenper Miller, John Morton, Dave Cowart, Dave White, George Alderman, Walt Bohren, Roger Mandeville, Buzz Marcus, Jim Downing and Amos Johnson. When the dust settled, Peter Gregg had won yet another Camel GT, ahead of Bill Whittington and John Paul, Sam Posey had won GTU over Dennis Aase and Don Devendorf, and Rob McFarlin had squeaked by Scott Hoerr and Gene Felton to win the Champion Challenge by less than 3 seconds.

July 4: SCCA Kendall Cup Nationals, Formula Atlantic and Vintage. Formula Atlantic came to Lime Rock fully formed; 1000 lb. cars carrying 220 hp Cosworth-Ford BDA twin-cams. There was a Labatt's Championship series, plus a $200,000 points fund sponsored by Ford of Canada. First at Lime Rock paid $10,000. Among the Atlantic stars—many of whom rose to international prominence—were Bobby Rahal, Danny Sullivan, Keke Rosberg, Howdy Holmes, Kevin Cogan, Price Cobb, Jeff Wood, Tom Gloy, Skeeter McKitterick, Chip Mead, Bobby Brown, Bertil Roos and Bill Brack. I count a future Formula One World Champion, two future IndyCar Champion/Indy 500 winners, at least three other Indy drivers and a couple of Can-Am stars. Yes, Virginia, you might say Formula Atlantic is a true training ground for professional racing drivers.

At Lime Rock that fateful day in 1978, the top 17 out of 30 starters broke the existing track record; Howdy Holmes was fastest with a 48.321 lap to grab the pole, then spun on the pace lap. He zipped through the grid on the next pace lap, led for 14 laps, broke the race track record with a 49.092 lap, then missed a shift coming through The Downhill. Bobby Rahal—already sponsored by Jim Trueman's Red Roof Inns—passed to win.

In the C-production race, Bob Leitzinger crashed into Paul Newman at the start, both in 280Zs. Leitzinger was out for the season with a broken collar bone; Newman started last on the restart, drove through the pack and finished second in what was really a memorable achievement.

August 27: First Annual Lime Rock Concours. Sponsored by Chinetti International Motors, 50 antique cars lined up in the Lime Rock paddock. Admission: $.99. Champagne lunch: $7.50.

September 4: SCCA New England Region Super National and Little LeMans for Showroom Stock Sedans. It was called Little LeMans, but this short Showroom Stock race had nothing to do with the 1957-'62 original.

(Left to right) Peter Gregg, Al Holbert, Lynn Griffis (Miss Camel GT) and Bob Sharp in the Lime Rock Victory Circle, 1974. The little boy is 1991 Trans-Am Champion Scott Sharp!

1979

March 31: VSCCA.

May 5: SCCA New England Region NARRC.

May 28: IMSA Winston GT, GTU, Champion Spark Plug Challenge and Kellygirl Challenge. John Bishop's "stock cars on road courses" series languished in 1977 and '78, but then picked up sponsorship from Kelly Services that included a $15,000 points fund to be shared by the three highest-placing women drivers. The cars turned out to be sort of like old-time Trans-Am sedans, but with 366 cubic inch V-8s in 3300 lb. cars. Among the contenders were Patrick Bedard in an Olds Cutlass built by Herb Adams, Gene Felton in a Buick Skylark, Lyn St. James in a Plymouth Volare, Jerry Thompson and Timmy Chitwood in Chevy Novas and Vern Smith in a Chevy Malibu. Gene Felton won in a Skylark from Pat Bedard in a Cutlass.

June 9: SCCA New York Region NARRC.

June 30: SCCA Northern New Jersey Region National. As of 6/30/1979, the Lime Rock lap records were:

AP: Panzarella, Corvette, 58.6
BP: Tullius, Jaguar, 59.2
CP: Sharp, Datsun, 59.6
DP: Catalano, Lotus 7, 1:00.2
EP: Dohner, Alfa, 1:01.4
FP: Newcombe, Spitfire, 1:00.2
GP: Wanta, Spitfire, 1:03.2
HP: Wentzell, Sprite, 1:04.0
ASedan: Posey, Mustang, 59.2
BSedan: Robinson, Datsun, 1:01.1
CSedan: Nagy, Mini-Cooper, 1:02.0
DSedan: Gross, Mini-Cooper, 1:07.4
Vintage: Finn, Cooper F/1, 1:06.1
GTX: Gregg, Porsche 935, 53.45
GTU: Posey, Datsun, 56.8
Kelly: Thompson, Nova, 59.52

FAtlantic and track record: Holmes, March, 49.09
FContinental: Whitney, Ralt, 54.8
Wrightson, GRD, 54.8
FV: Smith, Caldwell, 1:01.1
FF: Davey, Crossle, 56.1
S2000: Morton, Tiga, 59.4
ASR: Kovaleski, McLaren, 54.9
BSR: Cicale, Chevron, 53.8
CSR: Knoll, Bobsy, 55.9
DSR: Miller, Ocelot, Bobsy, 59.2
SSA: Ramirez, Porsche, 1:11.3
SSB: Kennedy, Alfa, 1:12.0
SSC: P. Hacker, Renault, 1:14.3
GTO: Currin, Corvette, 57.61
RS: Alderman, Datsun, 1:05.39

July 4: VSCCA and Lime Rock Concours.

July 21: SCCA New England Region NARRC.

July 28: SCCA Mo-Hud Region NARRC.

September 3: SCCA New England Region Super National, Little LeMans and Rabbit/Bilstein Cup.

October 13: SCCA New York Region NARRC Championship.

DAVID HOBBS

Has there ever been a better master of ceremonies than David Hobbs? The man is funny, erudite, vulgar and totally engaging, all the while bringing a wry and witty perspective to the subject at hand, whether it be a roast of Chris Economaki or an explanation of vintage racing to the audience of ESPN.

He is also a hell of a racing driver. David's thoroughly British father was Hobbs Transmissions, maker of, among other things, the patented Hobbs Mechamatic automatic transmission. Teenager David started racing with a Mechamatic-equipped Morris Oxford in 1959, graduating quickly to his father's XK-140. In 1961, he won 14 out of 22 races with a Lotus Elite fitted with a Mechamatic. The next year he raced an XKE and a Jag 3.8 sedan, plus won the Index of Efficiency at LeMans in an Elite.

David became a full-time professional driver in 1964, running Formula II for Merlyn, for Lotus in Cortinas and even a Spitfire at LeMans. In 1965, he moved up to a Lola T70, with which he did well both in Europe and America. That led to a berth with John Surtees' team, also in a T70, plus a Formula I ride with Tim Parnell. As Hobbs said at the time, "I've raced against World Champion Denis Hulme for years, and I was always as fast, maybe faster. That shows I should be a winner."

In 1967, Surtees and Hobbs had Aston Martin engines in their Lolas and failed to finish a race. The next year, in addition to a Honda for the Italian GP, he teamed with Paul Hawkins in a GT40 to lead Sebring, Daytona and LeMans before breaking each time. They did win at Monza and were second at Watkins Glen; Jackie Ickx and David won at Kyalami.

In 1969, Hobbs left semi-permanently for America. "The money is much better," he explained, "and the American racing drivers are not as good." He ran half the season of Formula 5000, yet won four races and was second in the championship. He was also third at LeMans with Mike Hailwood, in yet another GT40.

He won the Lime Rock Formula 5000 race in 1970—and broke the track record—then finished third in the series. He joined Penske Racing in 1971 to run USAC, plus ran F5000 for Lime Rock-based Hogan Racing. He dominated F5000, plus did well in USAC and the Can-Am. He eventually ran in four Indy 500s; his best finish was a fifth in a factory McLaren in 1974. After that, he linked up with BMW, driving a 3.0CSL at Daytona and Sebring, then a 320i in the Camel GT.

David Hobbs was too good to be a "journeyman" racer, but that's what he was...always professional, always competitive, but never getting the breaks that would see him at the very top of his profession. Like his friendly rival Sam Posey—with whom he shares many characteristics—David was already shifting from racing to television in the mid-Seventies, carving out another whole career that utilizes his vast knowledge of motor racing...and also his vast store of anecdotes. ■

David Hobbs isn't always the height of style, but he does bring, enthusiasm, excitement, humor and grace to everything he does from ESPN to an after-dinner speech, from Formula 5000 to IMSA to the Indy 500.

team rivaled Penske Racing in their cleanliness and degree of preparation. IMSA was top drawer in the Seventies. It didn't hurt that the most successful teams drove BMWs and Porsches. Socially acceptable drivers in Porsche Turbos is about as classy as you can get when it comes to racing. If it had been Ford versus Chevy, IMSA wouldn't have had the same appeal.

IMSA and Lime Rock

According to Jim Haynes, "When I started racing in 1957, John Bishop was the President of the SCCA. I had grown up racing on dirt tracks in the Midwest. At one of my first road races, at Watkins Glen, they made me start at the back of the pack because I was a novice. I passed about thirty cars in the first lap and when I came around again, I was second. John Bishop had me black-flagged. The next week he sent me a letter advising me that my style of driving was better suited to the professional ranks, and I should *not* compete in the SCCA!

"Well. I called up, and John and I eventually got to be very good friends. The SCCA headquarters was right in Westport, so we spent a lot of time together. When he left to form IMSA, he came to me and said, 'Jim, I need help. I need someplace to run my races.' So as one friend to another, I said 'Here's Lime Rock. Come run.' So we had one of the first IMSA races at Lime Rock and we had IMSA races every year after that. John and I are *still* personal friends, and Lime Rock is still the perfect place for IMSA."

It's a perfect match. Socially acceptable racers in fast, loud, expensive cars, racing at a picturesque venue that might have been designed for television coverage or a long weekend for Manhattan's Beautiful People. IMSA at Lime Rock—nothing else had quite the same subtle combination of qualities. That's what made it so important then and so memorable, today. In a very real sense, the history of Lime Rock in the Seventies is the history of John Bishop's IMSA, and vice-versa.

6. Amateurs in the Eighties

Production classes are the heart of the SCCA, stretching back in a direct line to the very foundations of American road racing. From one perspective, they haven't changed much in thirty-five years at Lime Rock; a Mini Cooper leads a Porsche 356 leads an MG-B. On the other hand, their classes are called GT5 and E-production, now, and they are real race cars clothed by replica vintage bodywork.

(Below) Current Lime Rock Vice-President of Operations and General Manager Mike Rand wheel-to-wheel with Lime Rock Vice-President of Marketing Phil Rosette during the Formula C Runoffs at Road Atlanta in 1980. Rosette finished second; Rand was fourth.

The big news at Lime Rock in the Eighties was that prior to the 1984 season, Jim Haynes arranged the sale of the race track from Harry Theodoracopulos—who'd owned it almost fifteen years—to a new group of investors formed by ex-Formula I driver and racing school owner Skip Barber. The price was rumored to be just over $1.2-million.

As Jim Haynes puts it, "Lime Rock ran in the black every year that I managed it for Harry, and I think he may have sold it for a million dollars more than he paid for it. Harry Theodoracopulos is probably the only person who ever made a penny out of Lime Rock Park!"

Skip Barber was actually the only one in the new group—they called themselves Lime Rock Associates, Inc.—who had any background in professional racing, but the others were—and are—all enthusiastic amateur racers. Lime Rock Associates includes six partners: managing partner Skip Barber; Robert Stocker, a Lansing, Michigan lawyer; Al Heinke, owner of a local construction company; Bill Bailey, owner of a marina in Florida; Fred Ruland, a Boston machine tool company executive and Paul Nachtway, a Boston lawyer. Jim Haynes stayed on as General Manager for two years.

New Look

Immediately, the new owners started improving the property. Nearly half the race track was resurfaced, plus a new paddock area was completed and new gas pumps installed. The hills were recontoured into better spectator viewing areas. That was just the first year. Lime Rock Associates has owned the track for nearly a decade, and for all that time they have pursued an aggressive improvement policy, investing over half-a-million dollars each year in capital improvements.

Among the most noticeable have been the Camel Bridge to the Outfield, connecting the Infield and Outfield for the first time since the track was opened. Over a period of eight years, three VIP Chalets were built, two on the Infield hill, one in the Outfield overlooking Big Bend, plus new restrooms in the paddock and the new building which houses

(Continued on page 151)

(Right) Formula Fords lapping Club Fords show the evolution of the species over the past twenty years. Lime Rock Park is almost ideally-suited for small formula cars.

NARRC

The North Atlantic Road Racing Championship was created by the ageless Henryk Szamota nearly thirty years ago. The NARRC concept is simple; a Regional level championship that encompasses all four local SCCA Regions, running events at a variety of area tracks. At the beginning in 1966, it was just for SCCA Area 1 drivers, now any SCCA Regional license holder can run the NARRC championship.

In terms of participation and purses—yes, purses!—the highpoint of NARRC so far was in the Eighties. Between 200 and 300 drivers would show up at twelve to sixteen races each summer. Lime Rock put up a purse of nearly $10,000; a separate drivers fund taken from the entry fees added an equal amount. Later on, Szamota obtained other corporate sponsorship to replace Lime Rock's contribution.

Lime Rock has always been the center of NARRC activities, because it's geographically in the center of the Regions involved. The NARRC Championships are always held at Lime Rock Park at the end of the season, too, and usually three or four of the dozen races that count as NARRC events are held here.

NARRC is SCCA amateur racing at its purest. Despite the introduction of purses there is very little money to win and little fame outside of a small circle of Regional enthusiasts. It costs almost as much to run for the NARRC Championship as it does to make a serious effort to win the Northeast Division Championship in Nationals, and it's almost as competitive, too.

How can that be? Because area stars like Peter Andrighetti, Ken Payson, Manny Matz, Doug Hobby and dozens of others run the same cars in Nationals and NARRC. People like Andrighetti even run Trans-Am with their "Regional" cars. Bottom line, NARRC is one of the most competitive race series in the country. And after three decades, it's still the personal project of the amazing Mr. Szamota.

Henryk Szamota was a bicycle racing champion in Europe before coming to this country. He joined the SCCA in 1951, and proudly carries SCCA member #67! He raced himself with an MG and a Jaguar XK-120M—which he bought in 1954 and still owns—but retired from driving in the mid-Fifties to concentrate on race organization.

NARRC founder Henryk Szamota, one of the most active members of the New York Region for over four decades.

He was Chief Steward at Lime Rock's Opening Day in 1957, was New York Regional Executive for years and then became Area 1 Governor for a record-setting two decades. He has been an SCCA Director on and off for years and is still incredibly active in Northeast Division events. He is regularly Chief Steward for Nationals and Regionals at Lime Rock, more than 40 years after he joined the SCCA. No one has been more active or more important to amateur racing in the Northeast. ■

1980

May 24-26: IMSA GT, GTU, Champion Spark Plug Challenge, Kelly Challenge, Vintage Invitational. The Camel GT race was run in two one-hour segments. In between, there was a one hour interval when, according to the regulations, "cars may be repaired and replenished freely." At the end of the hour, the racers were restarted in the order they finished the first segment. This obviated planned pit stops in Lime Rock's narrow pit lane. Among other novelties, John Paul, Sr. married Charlice Barnett—Sam Posey's mother, Justice of the Peace Mary Moore, officiating—outside the Hill Club tent. Later in the day, driving their Porsche 935 Turbo, John Paul, Sr. and Jr. became the first father/son co-drivers to win an IMSA GT race. Don Devendorf won GTU after Paul Newman, Sam Posey and Brad Frisselle ran into each other in Big Bend. Gene Felton ran away with the Kelly race in his Nova; Rob McFarlin edged Amos Johnson in RS.

July 3-5: SCCA Kendall Cup National, Formula Atlantic and VSCCA. Jeff Wood broke the track record with a 49.4 driving his Formula Atlantic Ralt. Among his Atlantic competitors at Lime Rock were Tom Gloy, Jacques Villeneuve, Peter Greenfield, Price Cobb, Whitney Ganz, Bruce MacInnes, Steve Saleen and Rickie Bell. At least a couple of these guys went on to be famous in places like Speedway, Indiana, and couple more you'd have to say are now infamous.

Among the other still-familiar names—but mostly in unfamiliar classes—were Rob Dyson, George Alderman, Bill Coykendall and Manny Matz in GT2, Doc Bundy in D-production, Louis Rettenmeier, Blair DuPont and Jim Briody in Formula Vee, Sam Gilliand in Sports 2000, Fred Wentzel in H-production, Joe Nastasi and Buddy Norton in Formula A, Roger Wiley in GT3, Chip Ganassi, Tom Davey, Drake Olson, Bob Herlin, Ken Duclos and Peter Klein in Formula Ford, Peter LoBianco in D-production, Bob Spreen and Alistair Oag in F-production, John Kelly in E-production, Paul Pettey in F-production, Peter Dus in B-production, Ken Slagle and Bob Leitzenger in C-production, Richard Abbate, Don Knowles and Stu Lasser in SSA, Steve Lewis and Dominic Cimino in SSB, Ken Payson and Johannes Krause in SSC. Many of these same drivers were running the NARRC series, as well.

July 6: NBA Bicycle Motocross. Among the leaders in the Northeast were Chuck Kurtz and Butch Leitzinger, sons of Sandra and Bob Leitzinger.

September 1: SCCA National, Rabbit/Bilstein Cup and Little LeMans.

October 11: SCCA New York Region NARRC Runoffs. *Road & Track* Nutworks Challenge. The first of John Dinkel's crazy handicap races, this one had long-suffering Sam Posey drive 5 laps in every competing car to set a "bogie" and the handicap start was figured so that theoretically, by lap 20, all the cars would be in the same spot at the same time. Obviously, it didn't work out that way, but then, that's what makes horse racing, as they say.

the motoring accessories shop called Driving Impressions.

Artist/racer Sam Posey and his half-brother, New Haven architect David Moore, have donated their time to design all these buildings constructed over the past decade, giving Lime Rock a unified look that it has never enjoyed before. Architecturally, Sam and David have attempted to combine traditional New England construction with the light, relaxed feeling of a nineteenth century seaside park. The atmosphere is meant to be traditional, but also functionally modern. Most of all, their style has given Lime Rock a festive air, totally in keeping with the "racing as family entertainment" that Lime Rock Park has come to personify.

Proliferation

Look at the SCCA racing classes for the Fifties. There were usually ten, and no one seems to have bothered to officially record lap speeds after the weekend unless somebody broke the absolute track record. Only from the early-Sixties on do we have updated lap records after each race. By then, the SCCA recognized 18 classes of racing cars.

With the introduction of SCCA professional racing in 1966 and the founding of IMSA in 1969, the great proliferation of amateur and professional racing classes began. By the Eighties, Lime Rock statistician Craig Robertson was keeping tabs on 34 separate classes. Today, it's up to 54 active classes, not counting 28 classes which are currently inactive and another 24 vintage and historic racing classes for which records are not kept!

An interesting quirk at Lime Rock is that from the very beginning in 1957, official lap records have always been set during a race, rather than in practice or qualifying. There have always been "unofficial track records" of sometimes questionable accuracy, but particularly in the early days of racing, there either was no timed qualifying or teams timed themselves. Accurate timing took place only during the race.

IMSA introduced the practice of publishing qualifying times, which at least in GTP, are often 2 seconds faster than the fastest race lap. So for the Eighties, not only are there over four dozen classes, some of them even have two lap records!

Is 54 too many race groups? Who knows. The good news is that there's a place in road racing for just about every car, every budget and every aspiration, whether you just want to

(Continued on page 155)

Lime Rock Expert John Sorbello gets a little wide in Big Bend in his C-sports/racing Sorbello during the New England Region National in 1989. C and D-sports/racing are really the last SCCA classes where an amateur driver/engineer can successfully design, build and race his own machine, relatively unfettered by rules.

1981

Kendall Refining Company built their second structure at Lime Rock, an 1100 square foot Press Building.

May 25: IMSA Camel GT, GTU, Kelly American, Champion Spark Plug Challenge. This was one of the classic years of IMSA; just about everybody who was anybody raced at Lime Rock this weekend. Among them were Al Unser, Jr.-Joe Crevier, Kenper Miller-David Cowart and David Hobbs in BMW M-1s, Bruce Leven-Hurley Haywood, the Pauls, Sr. and Jr., Gianpiero Moretti and Rolf Stommelen in Porsche Turbos, Klaus Ludwig in a Mustang Turbo, Brian Redman in the prototype GTP Lola T-600, Sam Posey, Don Devendorf and Paul Newman in Datsun ZX Turbos and Al Cosentino-Bob Speakman in an RX-7.

GTU was even more competitive, with Rob Dyson and Bill Coykendall in Datsun 200SXs, Logan Blackburn, Jim Fitzgerald, Bob Leitzinger, George Alderman and Bob Speakman in Datsun Zs, Walt Bohren, Jim Mullen, Al Cosentino, Jim Cook and Randy Pollack in RX-7s. Even Kelly American had such famous names as Cat Kizer, Lyn St. James, Vickie Smith, Amos Johnson, Tommy Riggins, Tim Chitwood, Robert Overby, Craig Carter, Rob McFarlin, Vern Smith and Gene Felton.

As of 6/30/1981, the Lime Rock lap records were:

Trans-Am: Holbert, Porsche Carrera, 57.10
BP: Dus, Corvette, 58.34
CP: Slagle, TR-8, 58.36
DP: Griffiths, TR-7, 1:01.15
EP: Dohner, Alfa, 1:01.40
FP: Newcombe, Spitfire, 1:00.20
GP: Wanta, Spitfire, 1:01.80
HP: Wentzell, Sprite, 1:03.50
GT1: Posey, Mustang, 59.20
GT2: Coykendall, Datsun, 1:01.07
GT3: Davy, Scirocco, 1:02.13
GT4: Foster, Civic, 1:05.33
F440: Tavolacci, Red Devil, 1:09.86
GTX: Hobbs, BMW M-1C, 52.22
GTU: Posey, Datsun, 56.80
Kelly: Smith, Camaro, 58.06
FAtlantic and track record: Woods, Ralt, 49.04
FContinental: Whitney, Ralt, 54.80
Wrightson, GRD, 54.80
FV: Smith, Caldwell, 1:01.10
FF: Ganassi, Crossle, 54.80
S2000: Morton, Tiga, 56.20
ASR: Sorbello, Sorbello, 53.80
BSR: Cicale, Chevron, 53.80
CSR: Knoll, Bobsy, 55.90
DSR: Miller, Ocelot, 56.85
SSA: Fazekas, Porsche, 1:10.07
SSB: Paxton, TR-7, 1:11.89
SSC: Salerno, Rabbit, 1:13.73
GTO: Cowart, BMW M-1, 56.49
RS: Archer, Renault, 1:04.63
Pro Rabbit: Benson, Rabbit, 1:11.6

July 4: SCCA National and CRC Trans-Am. Greg Pickett won the first Trans-Am at Lime Rock in six years.

September 7: SCCA Canon National, Rabbit/Bilstein Cup and Little LeMans.

October 10: Lime Rock 25th Anniversary Party.

SKIP BARBER

John "Skip" Barber is one of the few Americans to reach the top in professional racing—Formula I—then segue smoothly into a post-racing career. Skip started racing in 1958 when he was a student at Harvard; his first race at Lime Rock was a Regional in 1959 driving his Sprite. He tied for the F-production National Championship in 1962, then moved up through a Lotus 23 and a Brabham to a McLaren. He won three SCCA National Championships in a row with these sports/racers, then finished third in the 1967 USRRC.

Patron Gene Mason bought him a Tecno Formula B, followed by a Formula I March that was destroyed in Skip's first race with it. In a replacement March, Skip was 7th at Hockenheim, 14th at Zandvoort in Formula I in 1971. The next year, he was 11th at Watkins Glen in F/I, which led directly to a contract with March for Formula 5000 in 1973. As he says, "That car was a disaster." Almost overnight, a career that had taken fifteen years to build was over.

In 1975, he opened the Skip Barber School of High Performance Driving, now the largest driving school in the world, larger than all his competitors put together. In 1983, he put together his investors and made Harry Theodoracopulos an offer for Lime Rock Park.

Since then, he's started the Skip Barber Formula Ford Series and the Zerex Saab Pro Series, both intended to train young racing drivers to move up through the ranks. The list of his successful pupils reads like a Who's Who of Professional Racing. No matter how you view it, Skip Barber has had an incredibly varied and successful career. Lime Rock Park has been just one part—albeit an important part—of a life-long fascination with automobiles and motor racing. ■

(Lower left) Lime Rock Park President Skip Barber in his Formula B Tecno during the Lime Rock Continental race, August of 1969. (Below) Skip at the height of his career during a Formula I drivers meeting, 1971. (Bottom) Skip Barber School class at Lime Rock, 1984.

1982

In the fall of 1981, half the track was smoothed and repaved, mostly to silence critics who were complaining about the bumps. This was the first time the track surface had been redone in twenty-five years. In the spring of '82, in time for the Memorial Day IMSA race, artist John Peckham designed and painted the lime green LIME ROCK start/finish line logo on the new asphalt. Peckham's logo has subsequently appeared in thousands of photographs as an identifying feature of Lime Rock Park.

May 31: IMSA Camel GT, GTO/GTU, Kelly American, Champion Spark Plug Challenge, Renault Cup. Despite almost constant rain, John Fitzpatrick put his Porsche 935 on the pole with a 51.8 lap, then led flag to flag while setting a new IMSA Lime Rock lap record of 49.54. John Paul, Jr. was second in his Lola T-600 ahead of Ted Field, Danny Ongais and John Paul, Sr., all in Lolas. Don Devendorf in a Datsun won GTO from Gene Felton in a Camaro, Joe Varde in an RX-7 won GTU from George Alderman in a Datsun. Don Knowles won the Renault Cup race; Bobby Archer in a ProFormance Renault beat Amos Johnson in a Mazda in the Champion Spark Plug Challenge. Gene Felton beat Bob Overby in the American Challenge race, both in Camaros.

July 2-3: SCCA Kendall Cup National.

August 7: Northern New Jersey Region NARRC. Jim Haynes donated all the race tickets to the American Cancer Society, the Region donated all its proceeds and Connecticut Governor William O'Neill and Salisbury First Selectperson Charlotte Reid were on hand.

September 6: SCCA Canon Camera National, Bilstein Rabbit and Little LeMans. Among the winners were Ray Snowdon (F-production Spitfire), Bob Boig (G-production Fiat), Fred Wentzel (H-production Sprite), Jon Knight (GT4 Datsun), Perry Buzzell (Formula Vee), Steve Pieper (D-production Porsche), Tom Schweitz (GT3 Scirocco), John Kelly (E-production Elva), Drake Olson (Formula Ford), Eric Lang (Formula A March), David Ammen (Formula C Ralt), Steve Johnson (Sports 2000 Lola), Peter Dus (A-sports/racing Corvette), Ken Slagle (GT1 TR-8), Don Kearney (C-production Mazda), Tom Howen (GT2 BMW). In the Little LeMans race, Richard Abbate won SSA in his RX-7, Fran Dance-Bruce Kapstan won SSB with a Scirocco and Ken Payson-Kim Baker won SCC with a Fiat X1/9. Gary Benson beat the Hacker Brothers, Paul and Karl, in the Bilstein race.

October 9: SCCA NARRC Championship, *Road & Track* Nutworks Challenge.

have fun with your Showroom Stock commuter or whether you're grooming yourself for the World Championship.

The bad news is that even the most die-hard race fan has a hard time remembering the technical differences between Trans-Am cars and GTO, or who won which Championship last year. To non-racers, motorsports is so fragmented as to be incomprehensible, and that's not good.

At a cocktail party, when somebody says "Oh, you're a racing driver, didn't I see you crash on television in the Indy 500?" you can see their eyes glaze over as you start to tell them how you won the NARRC ITC Championship with your $1500 Fiat. It's not like baseball where there is one World Series, or football, where there is one Superbowl. One of the problems with having more racing classes than you can count is that everybody is a Champion of Something and people who aren't Racing People haven't a clue what you're talking about.

Part of this proliferation of classes—and consequent cocktail party confusion—is because between IMSA and SCCA, in the Eighties a concerted effort was made to please everybody...sponsors, manufacturers, career racers, club racers, *everybody*. Despite the cranky letters every month in *Sports Car*, the SCCA really did succeed in serving its member-

(Continued on page 163)

(Left) John Wall leads a rainy NARRC Showroom Stock race in 1990, driving his Mustang GT. It's painted white with blue stripes like an old-time Shelby Mustang. (Below) Peugeot's 505 Turbo was the hot set-up in Showroom Stock B during most of the Eighties, though the little Honda CRXs were the up-and-comers.

EXTRAORDINARY PEOPLE

For a while, the SCCA sold a bumper sticker that read "Ordinary people doing extraordinary things." When I saw one, I always thought—not of racing drivers—but of the zanies dressed in white who stand 20 feet away from the edge of a race track while barely under control cars careen toward them at 180 mph. For protection, they have some colored flags to hide behind. Extraordinary doings, indeed.

Who else would take off two vacation days from work, drive a couple of hundred miles, camp in a leaky tent, work from 8:00 am to 6:00 pm with only a short break for a hot dog and a warm Coke, discouraged by neither heat, nor rain nor storm nor gloom of night.

If they survive this ordeal, at the end of the day, their reward might be a free glass of beer, a shouting match with an irate racing driver pumped full of adrenalin, a long drive home in the darkness wearing grimy clothes and a splitting headache from too much sun. These people in white are not paid to do this, oh no. They *pay* for the privilege. Hell, they even buy their own uniforms and their own lunch. Extraordinary.

And they go to school. Your average person in white standing next to the Lime Rock track has not only graduated from an SCCA school for corner workers, but also knows everything from the proper etiquette to employ when passing a fellow competitor to the proper way to cut you out of a car with power tools and jump-start your heart.

At Lime Rock, there are 11 different flag stations and each station fulfills a dozen different functions. There is a flagman, responsible for flashing the yellow flag or oil flag to warn of danger, or the blue and yellow stripes of the overtaking flag. There are guidelines, but essentially the use of the flag is up to the discretion of the corner workers.

The phone operator is in constant communication with the Tower and other flag stations. If the communications go down, the race literally does not go on. There is a re-entry signaler. That's the worker wearing bright orange gloves who guides errant drivers back on to the course, not back into another advancing car. There are emergency men, people who push stalled cars to safety or run

(Above) The New England Region Grid Crew at Lime Rock, GTO/GTU weekend, September, 1991. Left to right, Karen Petersen, Sydnia Czarnacki, Laurie Sheppard, Kristen DuBois.
(Right) In a more casual age, corner workers watch for traffic while an OSCA vintage racer searches for electrical faults, 1974.

a hundred yards carrying a 50 lb. fire extinquisher into the teeth of a fire about to turn into an explosion. In charge of everybody is a station captain, who's keeping everybody else on their toes.

It takes over 100 people to properly conduct a race at Lime Rock Park, counting Registration, Scrutineering, Control, Flagging/Communications, Start/Finish, Stewards, Regional Reps, Pit and Paddock Patrol, Grid, Timing/Scoring and Fire/Rescue. About 40 percent of the workers are usually women, and most of them have been coming back to Lime Rock for years, perhaps decades.

Why? Well, a good number of workers are also racing drivers, and they like to feel they're giving something back to the sport. Another good-size group are people who'd like to be drivers, but haven't the time, the skill or more likely, a rich and generous uncle. Many workers are medical professionals who have a rare and useful skill that they know can be of benefit...emergency room doctors and nurses, for example, or volunteers from the local ambulance corps.

But most of the workers are car enthusiasts who got bored sitting on the hill watching the racers go round, and decided to get more involved. There's a tremendous feeling of camaraderie among the people in white, and many of them have become friends away from the track, as well.

Racing drivers have a special bond because of their shared experience; workers feel the same way. They find their own psychic rewards in not only being part of the show, but in making the show go on smoothly and safely. They even risk their lives to do it. Ordinary people doing extraordinary things? No way. These unselfish people are truly extraordinary. ■

(Left) A dead vintage racing Jaguar awaits a push from the corner workers.
(Below) Showing all the concentration of racing drivers, the corner crew before the Left-hander watches up and down the track, plus shows a yellow flag to approaching traffic.

Bendix
66
60
SCCA

SCCA
56
SCCA
FINMOR HOLDINGS

1984

Lime Rock Associates, Inc. bought Lime Rock Park from Harry Theodoracopulos. There were six partners; managing partner Skip Barber; Robert Stocker, a Lansing, Michigan lawyer; Al Heinke, owner of a local construction company; Bill Bailey, owner of a Florida marina; Fred Ruland, Boston machine tool company executive and Paul Nachtway, a Boston lawyer. Skip, of course, was a well-known professional racing driver; the others were all enthusiastic amateur racers he'd met through his driving schools. Immediately, the new owners started improving the property...nearly half the race track was resurfaced, plus a new paddock area was completed and new gas pumps installed. The hills were recontoured into better spectator viewing areas. In June, it rained so hard Big Bend was flooded deep enough for canoes to paddle around the course. Skip and his partners were probably wondering just what it was they'd gotten themselves in for.

May 26-28: IMSA Camel GT, GTO, GTU, Kelly American, Renault Cup, VSCCA. It rained and rained. Bill Whittington started on the pole with a 47.602 from his March/Chevy; Randy Lanier's fastest race lap was a 1:01.97 in a March/Chevy. Sarel Van Der Merwe won in his Porsche-powered March ahead of Lanier and Al Holbert, who had the first of the Porsche 962s. In GTO, Roger Mandeville's RX-7 beat Carson Baird's Datsun 300ZX. How hard was it raining? Walt Bohren qualified his Corvette on the GTO pole with a 54.046; his fastest race lap was a 1:09.63! Lime Rock Expert Paul Pettey exploited his local knowledge to finish sixth overall with his ex-Bob Tullius Jaguar XJ-S.

July 6-7: SCCA Kendall Cup National. Among the familiar names racing at this National were Kim Baker in an SSC X1/9. In Formula Ford, we find Frank Delvecchio in a Crossle, Jeff Andretti in a Swift FF and Paul Reisman in a Van Diemen. Omer Norton had his Austin-Cooper for GT5, Frank Righetti Sprites for both F and H-production, Randy Pollack and David Yarmoluk Mazda RX-7s for GT2 and Jim Fitzgerald the Bob Sharp Nissan for GT1. In Showroom Stock, Richard Abbate had his SSGT RX-7, Ken Payson his SSA Alfa GTV-6, Steve Lewis his SSA Toyota Supra and Mark Pocaro his SSGT Pontiac Trans Am.

August 3-4: SCCA Can-Am and NARRC Regional. Michael Roe set a new track record in practice at 44.61 in his Chevrolet/VDS. He also led every lap and won his fourth Can-Am in a row, beating Jim Crawford by nearly a lap.

August 31-September 2: SCCA National.

September 20-21: Vintage Fall Festival.

(Previous page) Spec Racers three abreast into Big Bend during the Trans-Am weekend, 1988. In many ways, this is the perfect amateur racing class...inexpensive to buy and maintain, free from depreciation and evenly-matched. How evenly-matched? At a typical Lime Rock Spec Racer National or Pro race, the first 15 cars on the grid will all have qualified within the same second!

ship with as many different classes as you can fit into a day of racing.

The Amateur Pyramid

Starting at the bottom of the SCCA pyramid, the Eighties brought us "Improved Touring," the *really* poor person's road racing class. The concept was simplicity itself. Take an obsolete Showroom Stock race car—which has no value to anybody—and "prodify" it the way we used to prepare cars in the Sixties for Production racing classes. To discourage big expenditures of time and money, restrict the races to Regionals only—you don't even get to go to Atlanta for The Runoffs, which is yet another incentive not to go crazy throwing money at your IT car.

How did it work? Well, there is obviously a market for inexpensive racing that is strictly a hobby; in less than eight years, IT has turned out to be even more popular than Showroom Stock, for the obvious reasons that the cars cost less, remain competitive virtually indefinitely and require just enough twiddling to make it interesting. There are no more satisfactory club racing classes in the world. Classes, indeed. From its humble beginnings, even IT has grown to five classes, if you count the new American Sedan Regional class for out-of-date V-8-powered Showroom Stock Camaros, Mustangs and the like.

Showroom Stock

Next up is Showroom Stock. IT has had the interesting effect of making a Showroom Stock car a better buy. One can acquire a Showroom Stock machine, race it in Regionals and Nationals for three years, in Regionals only for an additional two, then turn it into an IT car. The time limit on IT cars have yet to run out. Norman Latulippe, for example, has been running a Corvair in Northeast Division ITA races for years. At the front of the pack, too. The Showroom Stock cars of the Eighties are just now becoming eligible for IT...no doubt, they'll still be competitive twenty years from now.

Extended Lives

Extended competitiveness has actually been a problem in many SCCA classes. Even ten years ago, there were jokes that Production class grids and Vintage racing grids

(Continued on page 166)

(Left) Jim Crittenden at speed in the famous Mecca GT-1 Camaro in May, 1986.
(Below) Lorinda Cherry in her quick SSB Saab 900, July 3, 1982.
(Bottom) Norman Latulippe has been winning races at Lime Rock for decades in his Corvair, now running ITA.

1985

The first VIP chalet was completed and BMW was named the "Official Automobile" of Lime Rock Park.

May 24-27: IMSA Camel GTP, GTO/GTU, Kelly American, Champion Spark Plug, Renault Cup and the Great American Challenge Volley Ball Tournament. Local Boys Make Good: Drake Olson won in his first race with Rob Dyson's brand-new Porsche 962. In one of the best races ever at Lime Rock, Al Holbert's Porsche 962 led until blowing a pop-off valve, then Brian Redman in a Group 44 Jaguar and Olson had a tremendous battle until they collided, knocking the nose off Olson's Porsche. Olson then passed Redman in the rain, when the Jaguar slowed due to lack of visibility through the oil-smeared windshield. Redman drove around to the pits with the door open in order to see, but still finished second. A record crowd of 45,000 spectators were on their feet in excitement for most of the race. Father and son team Peter and Michael Greenfield won Camel Lights with their Tiga/Mazda. In GTO/GTU, Darin Brassfield won with the 7-Eleven Ford Thunderbird; Chris Cord was first in GTU in one of Dan Gurney's Toyota Celicas. Amos Johnson led the International Sedans race, ahead of his teammate Dennis Shaw. Craig Carter won the Kelly race and Lime Rock Expert Charlie Downes was the fastest of 39 Renault Encores.

July 4: IMSA Firehawk. Jon Milledge and Walt Maas won the Grand Sport race with their Carlsen Racing Porsche 944, lapping the entire field three times after 6 hours. Bob Henderson-Steve Lewis won the 3 hour Touring race in a Toyota MR-2. The Hacker Brothers were one and three in the 1 hour Compact race.

July 6: SCCA Can-Am and Kendall National. It rained and it rained and it rained. Crew members had to keep a hand on spare tires or they'd float down pit lane! Bruce MacInnes won the shortest Can-Am race in history, 26.01 miles in 16 minutes and 53.370 seconds.

August 10: SCCA NARRC, Trans-Am, Formula Atlantic and Sports Renault. Paul Miller in a Porsche Turbo Carrera beat Elliot Forbes-Robinson in a Buick by just .91 second. Wally Dallenbach, Jr. was third in a Capri. Paul Newman started from the pole and set a new Trans-Am lap record of 54.80 in his Nissan 300ZX, but he was in the pits two-thirds of the way through the race with a split radiator hose. Among the other winners were Bill Throop in a TR-4, Rob Sturgis in a Caldwell Formula Vee, Peter Andrigetti in a Corvette, Walter Nelson in a Lola Formula Ford, Michael Greenfield in a Ralt Formula Atlantic, David Murray in a Sports Renault and Bill Lockwood in a Showroom Stock Pontiac.

August 31: SCCA Playboy Endurance Cup.

September 2: SCCA Can-Am, National and Volkswagen Cup. Rick Miaskiewicz won the last Can-Am run at Lime Rock Park. Paul Hacker won the VW race.

September 20-21: BMW Vintage Fall Festival.

ROAD & TRACK MANUFACTURERS CHALLENGE CUP: IE. NUTWORKS

This zany day was John Dinkel's idea, back when he was editor of *Road & Track*. In the beginning, space salesmen, media buyers, PR guys and other friends of the magazine would each bring a new passenger car to Lime Rock on a beautiful morning at the height of fall foliage season. Dinkel would figure out a handicap using a complex system known only to him, and then everybody would race around for awhile. There was a backwards race, too, in which the competitors backed around the course. It was all good, clean fun.

By the end, racer/engineers like Chevrolet's John Heinricy were showing up with one-off cars built specifically for this event. In 1988, for example, Heinricy and Fred Schaafsma had an IROC Camaro with a 509 cubic inch big-block V-8. Not to be outdone, a couple of Pontiac engineers brought an aluminum-framed Fiero powered by an Oldsmobile Quad 4. The "guest drivers" were people like Irv Hoerr, Roger Marble, Tom Goad, Scott Harvey, Kjell Skavnes, Reeves Callaway, Mike Rand, Fred Stevenson, Bill Mayberry and Craig Horning.

The "sdrawkcab" race—started by Dinkel raising a checkered flag and ending when he raised the green—grew into an intense competition between local racers John Fitch and Sam Posey. It got to the point where General Motors actually built a couple of cars for Fitch and Posey that had the front suspension geometry reversed and the transmission gears diddled so that they were more stable—not to mention faster—backing up than going forward.

Wearing a tuxedo and tennis shoes, Sam zoomed around to set a backwards lap record of 1:56.630 in his Oldsmobile Custom Cruiser station wagon. That was nearly 4 seconds quicker than Fitch had gone the year before. The *R&T* Challenge lapsed after John Dinkel left the magazine, but it sure was great fun for the better part of a decade. ■

(Left) Our Gang—the racers, PR guys, ad salesmen and local hangers-on who attended the R&T *Challenge in 1987.*
(Below) Editor John Dinkel raises the green flag, signifying the end of the backwards race. Sam Posey is about to set a new backwards record of 1:56.630 in his "factory" Olds station wagon.

were indistinguishable; the same charge is even more appropriate today. This became a major issue with SCCA Club Racing during the Eighties; at what point do you decide everyone should be racing tube-frame Miatas rather than Porsche 356s?

The same impetus to reduce the costs of amateur racing that led to both cheaper series and long-lived eligibility of Production cars was at work in the Formula and Sports/racing ranks, as well. Club Ford, in particular, grew up to combat the "car of the month club" aspects of Formula Ford, while Spec Racer—nee Sports Renault—has addressed the same problem by rigid rules.

For amateur racers who just want to have fun, and who are paying their own bills, there is probably no smarter class than Spec Racer. The cars are evenly matched so the size of your racing budget matters only minimally, they are as inexpensive to maintain as a racing car can be, and they don't even depreciate because there is always another buyer ready to go racing on the cheap.You can even race in both Regionals and Nationals with a chance of winning. This is something that's difficult to do in other starter classes like IT or Showroom Stock. IT is strictly a Regional series; there is a vast difference between "Regional" and "National" cars, even in the "stock" classes. These problems don't exist in Spec Racer.

Lime Rock Amateurs

Amateur racing literally exploded during the Eighties at Lime Rock. There were Regionals and Nationals throughout the decade at which more than 300 racers showed up, straining the limits of the paddock, the grids and the time available, not to mention the tempers of the organizers.

As the Eighties went on, though, the dread "professionalism" reared its ugly head even at the Regional level. In the Seventies, heavily-endowed teams like Bob Sharp Racing or Group 44 were definitely the exception. In the Eighties, they might not have been the majority, but they were certainly no longer rare, even at the Regional level.

Turn-Key Racing

Probably the most significant trend in amateur racing during the Eighties was the rise of "turn-key" racing. No longer were amateur drivers willing to spend every evening and every weekend working on their race car. The Baby Boom generation wanted "instant gratification." Now! For the first time in racing history, large numbers of amateur drivers were willing to pay someone else to prepare their car and bring it to the track, so that all they had to do was show up and race. Or maybe, even more in the spirit of the Eighties, they'd just rent a racer when they felt in the mood, just as they'd lease their BMW, their condo at the beach or their furniture. The old bond between man and machine—formed during those long nights in a cold garage—was almost totally severed. This was a different kind of amateur racer than anyone had ever seen before.

It was amazing. Suddenly the Lime Rock paddock was filled with transporters that would turn Indy teams green with envy. Attended by uniformed mechanics, out would pop Spec Racers! IT Mazdas! Showroom Stock Mitsubishis! But the drivers were treated by their teams with all the pomp and circumstance that Penske Racing gives to Rick Mears.

So what if the driver was going to get a whopping bill on Monday morning, Friday and Saturday he was treated like a National Champion. Total professionalism had trickled down to even the starter levels of the sport. Road racing will never be the same.

7.
Modern Day Professionals

There's never been another decade like it. The staggering level of prosperity in America during the Eighties triggered an unprecedented deluge of advertising and promotional dollars, billions of which got diverted into professional road racing at every level.

(Previous page) The Camel Lights-winning Mazda/Denali of David Loring and Pierre Honegger, May 26, 1986. (Above) Bertil Roos in his potent Under 2.0-liter Can-Am Scania, at Lime Rock in 1983.

Corporate sponsors put their names on drivers, cars, individual events and entire series with casual nonchalance. Promoters stayed up nights inventing new types of racing; team managers went around with furrowed brows, thinking of ways to spend their sponsor's money. The number and variety and complexity of professional racing classes grew exponentially. It was a giddy time.

Can-Am

The SCCA let the Can-Am die after 1974, in favor of promoting Formula 5000 as its premier professional class. But Formula 5000, despite the fact that it was probably the best road racing series in the world at the time, never caught on. So in 1977, the SCCA changed its mind again and brought back the Can-Am.

Can-Am II or Son of Can-Am, if you prefer, was actually a pretty interesting series. The cars were made by simply putting sports/racer body shells over Formula 5000 single-seaters, which of course, made them even more aerodynamic. Not to mention, prettier. There was more room on the bodywork for sponsor decals, too.

Because Can-Am was an unlimited series—as long as you stayed under the 5.0-liter engine capacity limit—the first rebodied Lola F5000s soon gave way to purpose-built Can-Am cars with unfamiliar names like Frissbee or Invader. Herb Adams built one of the most innovative race cars ever, with the driver in one pontoon of the catamaran, the engine alongside him in the other.

That was just the kind of creativity an "unlimited" series is meant to foster. In addition, Tolemans and Ensigns—uncompetitive in Formula I—were given bodies and turned into stars in the colonials' Can-Am II. Later on, European Group C cars like Porsche's 956 ran in the Can-Am, as well.

Son of Can-Am accomplished a number of useful things. It gave the SCCA an exciting new class. It provided something to do with left-over Formula 5000 cars. It helped make stars of drivers like Geoff Brabham, Bobby Rahal, Danny Sullivan and Al Unser, Jr. It brought some wonderful racing to Lime Rock, starting with the 1983 season. But most of all, the Can-Am got John Bishop thinking about running similar cars under the IMSA banner. And thus was GTP born.

IMSA GTP

By far the most successful sports/racing series of modern times has been IMSA Grand Touring Prototype, which really got going in 1981. Prior to that, Bishop's top class was GTX (for "experimental"), which had become a one-design class of Porsche 935 coupes fitted with extreme bodywork both front and rear. In theory, GTX was for "production-derived" racers; in reality, the cars were one-off race cars suitable only for participating in IMSA GTX endurance races.

GTP was an acknowledgement by John Bishop that top-level racing cars need not be production derived, that there were FIA Group C cars that could be used for endurance racing in the States besides Sebring and Daytona, that the Can-Am II was being successful and IMSA had nothing like it.

Lola T-600

One car put GTP on the map. In 1980, Roy Woods and Ralph Kent Cooke put together Cooke-Woods Racing. Working with racer Bob Gar-

(Continued on page 172)

(Left) The Lime Rock Can-Am grid in 1984, Michael Roe on the pole with a new one-lap track record of 44.61 in his Chevrolet/VDS. Irishman Roe then went on to lead every lap and win his fourth Can-Am in a row. The single-seater VDS is probably the definitive design of the Can-Am revival; think of it as a stock-block Indy car with fenders.

1986

Jim Haynes resigned in February and in May, Jim Shane became the new Lime Rock Park General Manager. Most of the infield roads were paved and a new 220 foot skidpad circle was paved on top of the hill. A new .33 mile autocross track was built, as well. The second of the VIP chalets was also completed.

May 3: SCCA New York Region National. Among the highlights were Roger Wiley in a Datsun 510 zooming through the field from eighteenth to first in GT4; Tom Dolan barely edging out Wright Hugus and Jeff Solow in Formula Ford; Jim Crittenden in Mecca's Camaro beating Peter Andrighetti's Corvette in GT1.

May 26: IMSA Camel GT. This was the first year for Indy-style individual qualifying laps, but rain made a shambles of qualifying. Drake Olson was eventually given the pole in Rob Dyson's Porsche 962 even though Chip Robinson had set a new IMSA record in the Jaguar XJR-7 at 47.59. Olson led until giving over to co-driver Price Cobb. Driving solo and catching a lucky fuel stop under yellow, Al Holbert then went on to win in *his* 962 ahead of Cobb-Olson. David Loring-Pierre Honegger won Camel Lights in their Mazda/Denali. Irv Hoerr won his thrid straight Kelly race, with Patty Moise second. Al Salerno beat Paul Hacker in the ProStock Dash, both in VW Golfs. Tommy Archer won the ProFormance race in a Chevy Cavalier ahead of Dave Jolly's Mazda GLC.

August 9: IMSA Firestone Firehawk. Buddy Norton and Shawn Hendricks won with a Camaro ahead of Tom and Doug Goad in a Firebird. The Camaro was disqualified, however, and the Goads got the win.

August 16: SCCA Bendix Brake Trans-Am and Toyota Atlantic. The only Lime Rock Trans-Am winner to get a color photo on the cover of *USA Today*, Paul Newman outlasted Wally Dallenbach and Pete Halsmer, outlucked Scott Pruett and Chris Kneifel and outraced Elliot Forbes-Robinson to win in Bob Sharp's Nissan 300ZX Turbo.

August 30: SCCA New England Region National, VW Cup and Lucas Sport Renault. Among the well-known names in Lucas Sport Renault were winner Scott Lagasse and runner-up Ben Burrell, Dorsey Schroeder, Robbie Buhl, Todd Morici, Ray Kong, Jeff Purner, Lisle Gilbert and Richard LaPierre. Al Salerno won the VW Cup from Peter Schwartzott and Alistair Oag. Among the National racers were Fred Wentzell, Bob Boig, Paul Nachtway, Peter Klein, David Thompson, Paul Newman, Peter Dus, Jim Crittenden, Scott Sharp, Scott Quaile, Jim Fitzgerald, David Yarmoluk, Peter Andrighetti, Bob Earl, Tommy Kendall, Chris Schneider, Tom Dolan, John Gimbel, Wright Hugus, Frank Del Vecchio, Mark Pocaro, Jerry Lustig, Garth Ullom, Stuart Lasser, Chuck Cunningham and Ken Payson. Yes, you might say Northeast Division Nationals are competitive!

September 1: IMSA GTO/GTU. Jack Baldwin set a new GTO lap record in qualifying at 52.404, then led almost every lap to win GTO in his Peerless Racing Camaro. Tommy Kendall, driving the five-year-old ex-Downing, ex-Baldwin Mazda RX-7, was fourth overall, first in GTU.

September 12-14: BMW Vintage Fall Festival.

1986 Records

Class lap records at Lime Rock, as of 10/16/86:

Vintage: Akin, Cooper Monaco, 1:02.32
Historic: Buzzetta, Lola T-292, 54.98
BP: Dus, Corvette, 58.34
CP: Slagle, Triumph TR-8, 58.34
DP: Poduje, Lotus Elan, 59.92
EP: Kelly, Elva Courier, 1:00.51
FP: Newcombe, Spitfire, 1:00.20
GP: Boig, Fiat X1/9, 1:01.19
HP: Wentzell, Sprite, 1:01.31
GT1: Dus, Corvette, 54.53
GT2: Sharp, Nissan, 57.51
GT3: Schmid, RX-7, 59.08
GT4: Schweitz, VW Rabbit, 1:01.77
GT5: Firlein, Datsun B210, 1:03.28
SSGT: Baker, Corvette, 1:04.94
SSA: Ullom, Dodge Shelby, 1:06.93
SSB: Lewis, Toyota MR-2, 1:08.10
SSC: Miller, Honda CRX, 1:09.64
IT Enduro: Payson, Alfa, 1:07.96
ITS: Pollack, Mazda RX-7, 1:07.14
ITA: Latulippe, Corvair, 1:07.10
ITB: Hunt, Volvo 142, 1:07.27
ITC: Stidsen, Fiat Strada, 1:12.4
ASR: Sorbello, Sorbello, 53.80
CSR: Thompson, Ralt RT-5, 53.57
DSR: Miller, Ocelot, 56.85
S2000: Johnson, JRE SR71, 54.38
SRenault: Lagasse, Sports Renault, 1:01.39
FA: Fish, Ralt RT-4, 48.40
FC: Rettenmeier, Ralt RT-4, 51.07
FF: Rubio, Swift DB-1, 53.83
Club Ford: Mann, Van Dieman, 56.67
FV: Buzzell, Zink V4, 1:00.76
F440: Drissel, Red Devil, 59.65
Barber Saab: Unser, Saab, 57.00
Can-Am and outright track record: Jacques Villeneuve, GR-3, 46.93
Can-Am, Under 2.0-liter: Greenfield, Ralt, 48.81
Trans-Am: Dallenbach, Jr., Camaro, 52.76
Pro Rabbit: Hacker, VW Rabbit, 1:08.14
VW Golf: Oag, VW Golf, 1:08.77
GTP: Olson, Porsche 962, 47.47
Lights: Alsup, Buick/Royale, 52.87
GTX: Fitzpatrick, Porsche 935, 49.54
GTO: Brassfield, Ford Thunderbird, 53.85
GTU: Hayge, Toyota Celica, 56.26
IS: Shaw, Mazda GLC, 1:03.43
AC: Riggins, Buick Somerset, 56.14
ProStock: Pobst, VW Golf, 1:10.25
Renault Cup: Knowles, Renault LeCar, 1:12.32
Renault Cup: Travis, Renault Encore, 1:11.37
Firehawk GS: Norton, Camaro, 1:07.10
Firehawk Sports: Tosi, Shelby Charger, 1:07.95
Firehawk Touring: Hacker, MR-2, 1:09.01
Firehawk Compact: Hacker, VW Golf, 1:10.87

Jacques Villeneuve, outright track record holder in 1986. His record was 2 seconds slower than Michael Roe's one-lap record set in a 1984 practice session, but at Lime Rock, only lap times set during a race are considered for official lap records. Unofficially, David Hobbs had gone faster than Villeneuve fifteen years before in a special Formula 5000 car with an engine that grenaded after two laps.

(Above, left to right) Brian Redman and Derek Bell have been friendly rivals in international endurance racing for decades. (Right) Bob Tullius took Jaguar back into endurance racing in 1982 for the first time in twenty-five years with his Lee Dykstra-designed XJR-5 for IMSA GTP.

retson, Lola importer Carl Haas, and Eric Broadley of Lola, they financed a Chevy-powered Lola prototype coupe. It was one of the first race cars in the world to achieve "ground effects" without the use of side skirts.

With Brian Redman driving, Lola T-600 Serial #1 won its first three races in 1981 (including Lime Rock) and eventually finished the season with five firsts and five seconds out of ten races. Of course, Brian Redman and the Lola T-600 won the 1981 IMSA GT Championship. The Lola was dazzlingly pretty, reliable, easy-to-drive and fast. Right out of the box, it was a winner. Within a year or two, everybody in IMSA had a Lola T-600, most with Chevy V-8s., though Gianpiero Moretti did have one with a Porsche engine.

Perfect Timing

As the Reagan economy took off, so did GTP...like a skyrocket. In 1982, Group 44 showed up with their Jaguar XJR-5, by far the most ambitious racing program that Bob Tullius had ever attempted. He kept improving both the car and the team for the next five years.

BMW had its M-1C, then March came out with its 83-G coupe that could be powered by a turbo BMW or Porsche engine. There were more Lola T-600s, the front-engined Ford Mustang GTP, Gianpiero Moretti's one-off Momo/Ford powered by a 3.0-liter Cosworth. To be competitive in the early-Eighties, you needed around 650 hp in a car than weighed less than 2000 lbs.

As it turned out, you also needed a Porsche 962. The 962 first appeared in IMSA in 1984, and everyone else might just as well have stayed home. Al Holbert and Bob Akin had the first two seen at Lime Rock. Holbert was third behind two Marches, but the message was clear. The Porsche 962 would come to dominate IMSA and Group C in the Eighties as thoroughly as the fabled Porsche 917 had dominated Can-Am and Group C in the Seventies.

How long did this domination last? As Skip Barber says, "The highpoint of IMSA GTP was probably 1988. After that, the effects of the Recession started to make themselves felt." A

(Continued on page 176)

AL HOLBERT

(Below, left) Al Holbert and Pam Meadows, Miss Camel GT, on the Lime Rock victory stand after the Camel GT, May 26, 1986.
(Below) Holbert's Lowenbrau Porsche 962 on the way to victory earlier that afternoon. These 962s were almost unbeatable in the mid-Eighties, thanks to a typically-Porsche combination of speed and reliability...and a typically-Holbert level of race car engineering, preparation and race day strategy.

Al Holbert grew up around road racing; his father, Porsche dealer Bob Holbert, was one of the best known stars of the Fifties and Sixties. Young Al started racing in 1971 with a Porsche 914/6. He won his first race. Two years later, he was second in the Under 2.5-liter Trans-Am Championship, in 1974 he won his first professional race, the IMSA Camel GT at Road Atlanta. By 1976, he was already at the top of his profession, winning the Camel GT Championship in his Chevy Monza and being invited to run in Roger Penske's IROC races.

For the next twelve years, Al Holbert won an astonishing number of races. In his total career, he won LeMans three times—in 1983, '86 and '87—and the Daytona 24 Hours twice. With a total of 63 wins in major professional races, including 49 in IMSA Camel GT, 10 in the Can-Am and one in Trans-Am, he was far and away the most successful road racer of his era.

Even more impressive, he ran his own team and designed some of his own winning cars. As his long-time co-driver Derek Bell once said, "Al Holbert was the greatest engineer of a racing car that I have ever known." He was killed on September 30, 1988 when his light plane crashed on take-off.

At the memorial service held before the Columbus 500 race in 1988, Derek Bell said, "Al helped a lot of people, people who were perhaps destitute as mechanics. He would pick them up off the ground and give them confidence in themselves." Most race fans only knew that Al Holbert could drive extremely quickly around a race track, and that Lime Rock was one of his favorite places. His friends knew that he was a wonderful human being as well, the nicest man you'd ever want to meet. ■

00

1988

At Lime Rock Park, a new high-volume water supply was installed and more infield roads paved.

May 7: SCCA New England Region National. Jerry Hinkle started last in the small-bore production race, swept through the field, lowered his class lap record to 1:01 and won overall. In GT1, Peter Andrighetti, Rickie Bell and Paul Miller put on quite a show. Miller's Porsche eventually won when Bell overcooked the tires on his Camaro and Andrighetti broke a rear suspension arm.

May 30: IMSA Camel GT, Barber Saab, International Sedans, American Challenge, PaineWebber Vintage Invitational. On the Tuesday practice day before the race, Geoff Brabham in the Nissan GTP turned an unofficial 45.7 lap, a full second quicker than Price Cobb's track record. In practice, Martin Brundle lowered the track record to 44.885 with his Jaguar and started on the pole for the race. Brabham's teammate John Morton got airborne at the top of the hill and destroyed his Nissan, luckily with almost no damage to himself. In the race, Brabham drove solo to win ahead of John Nielsen-Martin Brundle in the Jaguar XJR-9. Brabham's Lime Rock victory was the third of eight straight and nine total, which gave him the IMSA Camel GT Championship. Tom Dolan won the Barber Saab race; Dick Danielson won over Clay Young in AC. Amos Johnson beat Parker Johnstone in IS.

July 4: SCCA Kendall Cup National. Among the winners were Dave Drissel in Formula Vee, Soter Slomski in D-sports/racing, Hamilton Shepley in Sports Renault, Chris Fahan in Formula Ford, Scott Sharp in GT1, Ken Caselnova in GT2 and Ken Payson in SSGT/SSA.

August 6: SCCA Trans-Am, Formula Atlantic and Sports Truck Challenge. Scott Pruett led from wire-to-wire to win the Trans-Am in his Merkur XR4Ti; Darin Brassfield was a close second. This snapped a string of Trans-Am victories by the Group 44 four-wheel drive Audis driven by Hurley Haywood, Walter Roehrl and Hans Stuck (who wasn't at Lime Rock). Haywood and Roehrl were fourth and fifth at Lime Rock. In the first SCCA Racetrucks race at Lime Rock, Ray King and Jeff Krosnoff in Nissans beat Tommy Archer's Jeep. Freddie Rhemrev won the Formula Atlantic race.

September 3-5: IMSA Camel GTO/GTU and Firestone Firehawk. Scott Pruett got all the headlines by racing in the CART, Trans-Am and IMSA GTO races, all in one weekend, at Mid-Ohio and Lime Rock. He started from the pole in the Trans-Am, second behind teammate Halsmer at Lime Rock in GTO, eighth on the grid in the CART race. Then it all went to hell. Faulty windshield wipers put him into the pits during the Trans-Am. He finished sixteenth. He crashed in the CART race and finished twentieth. At Lime Rock, a rocker arm broke in his Merkur after only 3 laps. Despite this dreadful weekend, a month later, Truesports picked him to replace Bobby Rahal in their Indy car. Wally Dallenbach, Jr. won the GTO race. Local star Buddy Norton teamed with John Stump to win the Firestone Firehawk GS race. Lance Stewart co-drove in different Hondas with Parker Johnstone and P.D. Cunningham. Stewart finished first *and* second in Sports.

(Previous page) May 31, 1982: Danny Ongais in Ted Field's Interscope-sponsored, Chevrolet-powered Lola T-600. These Lolas, particularly in Field's classy black livery, were just about the prettiest racing cars of modern times. They were also incredibly successful; Brian Redman won five out of ten races in his Lola T-600 in 1981, and finished second in the other five to win the IMSA Camel GT Championship.

crowds were bigger, the prestige was greater. If you were a team looking to go racing, there was really no choice.

Trans-Am never went away, it just sort of staggered along throughout the Seventies, some years as a single class, some years as two classes for big and small cars. Lime Rock, like many another American race track, dropped the Trans-Am after 1974 because it just wasn't a viable series.

The explosion of racing in the Eighties changed all that. Automobile companies were desperate to win *some* championship they could tout in their advertising, and Trans-Am did have priceless name recognition. Did Pontiac have a model called Camel GT? Of course not.

The Trans-Am was restructured once again in 1980, and this time there was no nonsense about Under 2.0-liter or anything else. There was one class, a weight/displacement formula to even out the competition and a series sponsor in Bendix. The Trans-Am came back to Lime Rock again in 1985.

While the SCCA was fine-tuning Trans-Am, of course, IMSA had been running GTO races since the very beginning, races for cars that were built to essentially the same Trans-Am rules. For the first decade, GTO ran as the "slow class" in the same races with the fast Camel GT cars. With the introduction of the incredible GTP machines, however, at some tracks—like Lime Rock—IMSA wisely put GTO into the same race with GTU, leaving GTP and the new Camel Lights together as the Camel GT. This was a much better match all-around, and also gave IMSA two major IMSA road races at Lime Rock to

(Continued on page 184)

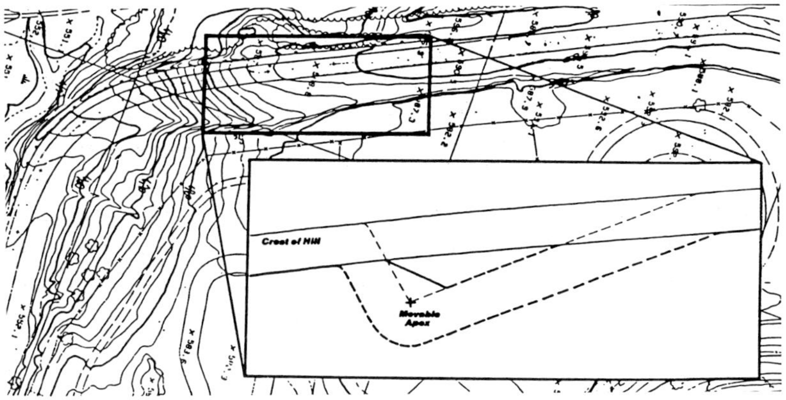

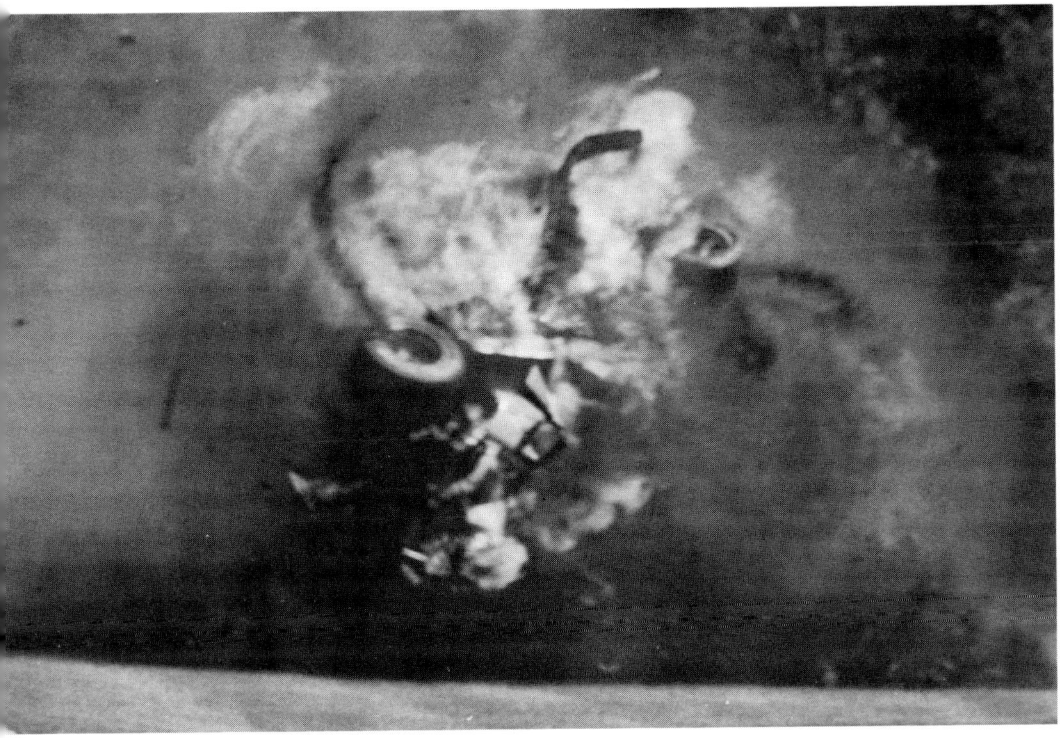

(Left) John Morton miraculously survived this practice crash of his Nissan GTP car in 1988, after it became airborne at the top of The Uphill. In a show of soul-searing bravery, Geoff Brabham took his identical Nissan and won the race two days later, knowing the same thing could happen to him.
(Above) Before the 1989 IMSA race, the "John Morton Chicane" course extension was installed at the top of the hill, forcing GTP cars to brake, rather than accelerate, at that point.

1989

May 6: SCCA New England Region National. Among the racers were three defending National Champions—Scott Sharp in GT1, Dave Drissel in Formula 440 and John Thompson in Formula Atlantic.

May 29: IMSA Toyota Trucks Grand Prix, LuK Clutch International Sedans, TDK Barber Saab and PaineWebber Vintage Invitational. This was the first year for the John Morton Chicane at the top of The Uphill, designed to force the IMSA GTP cars to brake at the top of the hill rather than accelerating and becoming airborne as John Morton's Nissan had in 1988. The new "long course" was 1.54 miles, compared to the traditional "kinkless" course of 1.53 miles.

Lime Rock Expert Drake Olson started from the pole and led the Camel GT race until forced by IMSA officials to stop to fix an inoperative taillight. Geoff Brabham then passed and went on to win with his Nissan GTP-ZXT. Olson's Toyota Group C quit with an engine problem and did not finish. Jan Lammers was second in the Jaguar XJR-10; Price Cobb-John Nielsen were third in the Jaguar XJR-9. Brabham won $113,000. Charlie Morgan won Camel Lights. Claude Bourbonnais beat Colin Trueman in the Formula Atlantic race, both in Swifts. Amos Johnson, Dave Jolly and Dennis Shaw, all driving Mazda MX-6s were 1-2-3 in the International Sedans race. Johnson won $6300. Bernard Santal, Justin Bell and Robbie Buhl finished in that order in the Barber Saab event. Santal won $8500.

This was also the beginning of the Toyota Trucks sponsorship of Lime Rock Park. Two Toyota V-6 trucks were fitted with Century wrecker bodies; two were equipped as "immediate response" fire trucks and four more as tow/track maintenance vehicles.

July 1: SCCA Nationals. Local hero Rickie Bell, driving a $250,000 Camaro built by Mecca Development of nearby Sharon, won his fifth straight GT1 National. He qualified at a 53.96, well under the GT1 lap record. Two weeks later at Watkins Glen, Bell wrapped up the Northeast Division Championship. Not only had he won every race, but started on the pole and led every lap. At The Runoffs, Bell finished third.

July 4: IMSA Firestone Firehawk Endurance. Peter Cunningham won in his Camaro ahead of the Camaro of Mitch Wright-Kim Baker and the BMW M-3 of Ray Korman-Ron Christensen. Leaders Doug Goad and Cunningham collided going into Big Bend; P.D. went on to win, Goad limped in on a broken wheel.

August 5: SCCA Trans-Am and Escort. There was a six-car crash on the first lap of the Trans-Am caused by Paul Gentilozzi; front-runners Paul Newman, Rickie Bell and Lyn St. James were casualties. Skip Barber instructor Dorsey Schroeder won with his Mustang. John Heinricy-Don Knowles won their eighth straight Escort race in a Camaro; Dorsey Schroeder-Paul Dallenbach were second in a Saleen Mustang.

September 4: BMW Vintage Fall Festival and 1959 Formula Libre Reunion.

September 30: IMSA GTO/GTU/AC and Sisapa Pro Sports 2000. Hans Stuck won from Pete Halsmer in GTO; Tom Jagemann won S2000 in his Lola.

(Left) Lime Rock Trans-Am, 1989. Scott Sharp is the meat in the sandwich of Tommy Kendall and Paul Gentilozzi.
(Large photo) Pete Halsmer beat teammate Scott Pruett by 4 seconds to win the 1987 Lime Rock Trans-Am. They drove matching Merkur XR4Tis.
(Bottom) Prepared by Group 44, Audi's unique four-wheel drive Trans-Am sedans won five races in a row in 1988. The SCCA rewarded them by making each car carry an extra 200 lbs. starting with the Lime Rock event. Hurley Haywood got this Audi into fourth; Walter Roehrl was fifth. Scott Pruett won with his Merkur.

cars. Only a factory-backed team had a chance, and even then the competition was fierce.

At lower levels, the costs of racing had also gotten out of hand. To expect to win any national level professional series, even in Showroom Stock, one needed factory support and a six-figure budget. Endurance racing, which gained in popularity in the Eighties, was even worse. For one thing, it literally *consumed* cars; top-rank teams built a new car for every long race. This was not something you could do in your backyard.

It is well-documented that during the Reagan Presidency, the economic spread between poor people and rich people increased dramatically. The same thing is true of road racing. The costs of amateur racing escalated, it's true, but nowhere near as much as the costs of professional racing. The poor had to be richer and the rich had to be even richer. It was a very unusual time.

Golden Era

I think we'll look back and see the Eighties as a Golden Era in American professional road racing, as a time when cubic money rather than cubic inches made the cars go fast. Surely Al Holbert in a Porsche 962 or Geoff Brabham in a Nissan GTP ZX will be remembered as fondly as Mark Donohue in a Porsche 917-30 or Denis Hulme in a McLaren M-8A.

Lime Rock was in the center of everything. At the high-point, say 1986 through 1989, Lime Rock hosted all the IMSA classes at least once during the season. We had GTP, GTO, GTU, the LuK Clutch International Sedans, Kelly American, Firestone Firehawk Endurance and Barber Saab Pro Series.

We had all the SCCA professional series, too...Can-Am, Trans-Am, Toyota Atlantic, Sports 2000, Super Vee, Pro Sports Renault, Volkswagen Cup and even the silliest road racing idea since Patrick Jacquemart thought of racing LeCars, the Coors RaceTruck Challenge for pickup trucks.

There was so much money around, you could sell sponsorship for *anything* in the Eighties, from GTP cars that cost a million dollars to pickup trucks that weren't worth two cents. It was a boom time, a grand time, a time of fun and games and serious racing on Other People's Money.

To see the Lime Rock spectator hill covered with two dozen corporate hospitality tents and 40,000 fans cheering on stars of the caliber of Al Holbert and Pete Halsmer, Scott Pruett and Chip Robinson, Bob Dyson and Price Cobb, Hurley Haywood and John Morton, was to see professional motor racing at its best. Twenty years from now kids will say "You were at Lime Rock for the 1987 IMSA race? Wow! That must have been *neat.*"

8.

Lime Rock Today and Tomorrow

1990

In the Spring, Lime Rock Park General Manager Jim Shane moved back to California and was replaced by David Hall, Vice-President of Operations and Mike Joy, Vice-President, Marketing and Communications.

May 5: SCCA New England Region National. Among the winners were Doug Hobby (FF Reynard), Rickie Bell (GT1 Camaro), Don Meluzio (GT3 Porsche 924), David Lapham (SSA Mitsubishi) and Ed Breault (SR).

May 25-28: IMSA Camel GT, Barber Saab, Toyota Formula Atlantic, LuK Clutch International Sedans and PaineWebber Vintage Invitational. Geoff Brabham, Kevin Cogan, Rocky Moran and John Paul, Jr. all raced in the Indy 500 on Sunday, then flew back to race in the Camel GT on Monday. The best finish out of the four at Indy was Cogan, who was ninth. At Lime Rock, Drake Olson put his Toyota on the pole with a track record 47.078, but Price Cobb-John Nielsen won the Camel GT race with the Jaguar XJR-10 ahead of Wayne Taylor-Jim Miller in a Chevrolet/Spice. Of the Indy racers, Geoff Brabham was fourth, Rocky Moran fifth, Kevin Cogan and John Paul, Jr. shared the Nissan in thirteenth. Martino Finotto-Ruggero Melgrati won Lights with their Ferrari/Spice. Parker Johnstone won the International Sedan race with his Acura Integra over Scott Hoerr in an Oldsmobile Calais; Rob Wilson won Barber Saab over Walt Bohren; Claude Bourbonnais won Toyota Atlantic for the second year in a row. He also set a new record at 56.98.

June 2: SCCA New York Region NARRC.

June 30: SCCA Northern New Jersey Region National. Among the winners were Keith Drissel (F440 Raptor), Alex Miller (CSR), Jerry Randall (S2000 Lola), Steve Lewis (SSC Toyota MR-2) and Eric Krueger (EP MG-B).

July 4: EMRA Independence 200. The star was Andy Murray, driving a '64 Malibu, who cut consistent 1:03s!

July 21: SCCA New England Region NARRC.

August 3-4: SCCA NARRC and Sisapa Pro Sports 2000. John Fergus won the Sisapa race.

August 18: SCCA Mo-Hud Region NARRC.

August 25: SCCA New York Region Endurance Race. The overall winners were Paul and John Reisman in an ITE Pontiac Firebird. Cooper-Lynch won SSGT with a Mustang, David Brand drove solo to win ITS with a Mazda RX-7, Jack Keller drove solo to win SSA with a Dodge Shelby, Korinis-Mina were first in ITA with a BMW and Green-Bucher were first in ITB with a similar BMW. Joshua Wortzel won SSB with an MR-2. ITC was won by Jeff Haines in a Rabbit; SSC by O'Connor-Daly in a Golf.

August 31-September 3: BMW Vintage Fall Festival and 1970 Trans-Am Reunion.

September 22: SCCA New York Region NARRC Runoffs.

September 28-29: IMSA GTO/GTU. Robby Gordon in a Mercury Cougar won overall and GTO, from Steve Millen and Jeremy Dale, both in Nissans, and Dorsey Schroeder in the other Cougar. Don Knowles was fifth overall, first in GTU, with his Dodge Daytona. Dan Osterholt won the All American Challenge class.

On October 18, Phil Rosette replaced Mike Joy and was named Vice-President, Marketing and Promotions.

rate cars for Indy and Formula I?"

There will always be racing, in automobiles as long as there are automobiles, and in something else when cars are obsolete. Call it a socially acceptable substitute for war, a way to get your kicks, an adrenalin rush, whatever. Human beings have always raced something, and they always will. No doubt, there will always be a sponsor willing to paint his name on the side, whether it's a Roman chariot, a Sports 2000 or hydrogen-powered transportation module. That's not the question. The question is, What form will this racing take?

Spec Racing

The answer? One-design racing. Think about it. Among today's most successful road racing classes are Spec Racer, Barber Saab, Indy Lights, Barber School Formula Fords and when it gets going, perhaps the Shelby Can-Am. These are all one-design classes in which everyone uses the same car outfitted the same way.

Even professional racing classes like NASCAR, Trans-Am and GTO use the same tires, the same engines and usually, the same chassis for everyone. This is true even for IndyCars, where everyone uses the same tires, similar engines and chassis that are almost identical because they've often been designed by the same people at different stages in their careers. Of course, there are also events like IROC, where the cars are as evenly matched as humanly possible.

Turn-Key or One-Design racing is the coming thing, because it puts the emphasis on the driver rather than the car. In the early days of motoring, manufacturers felt it necessary to prove the reliability and performance of their cars. Today, every new car consumer expects that his car will successfully traverse 100,000 miles without opening the hood in anger.

With the notable exceptions of IMSA GTP and Formula I, and to a lesser extent, Trans-Am and GTO, road racing in the Nineties is mostly about the individual performance of the drivers, about who is more skillful, more daring, more experienced. It has become a test of man, not machine, and that is wonderful from several points of view.

One, it makes racing more "pure," in the sense of what politicians call "a level playing field." No driver possesses an "unfair advantage." Two, it makes a better show, because as any race promoter knows, the crowds come to see competition between personalities, not technical sophistication under the hood. Three, it allows drivers who really do have superior talent to strut their stuff, hopefully getting recognized by The Right People and moved up the racing ladder to the top.

There's another factor at work here, too. Modern racing cars have become so complicated, even on the Showroom Stock level, that only trained racing mechanics really understand how to set them up correctly. The "turn-key" race car, tuned and maintained by experts, is already a fact of life. The "shade tree tuner,"

(Continued on page 197)

Install a turbocharged Saab engine in a state-of-the-art formula car chassis and you get Barber Saab, a potent machine capable of lapping Lime Rock's IMSA "long course" in 54 seconds. Make all the cars as identical as humanly possible, and you get great "one-design" racing that's the perfect training ground for young professionals on the way up.

TOYOTA
98

98
GOODYEAR

1991

May 4: SCCA New England Region National. Among the well-known race winners this weekend were John Wall (SSGT Camaro), Ken Payson (SSB Peugeot), Bob Boig (FP Bertone), John Sorbello (CSR Swift DB-2), Scott Quaile (GT1 Camaro), David Pettigrew (GT3 Mazda), Doug Hobby (FF Swift) and Jim Johnston (FV Caldwell).

May 24-27: IMSA Toyota Trucks Camel GT, Potenza Supercars, Toyota Atlantic, Barber Saab, International Sedans and PaineWebber Vintage Invitational. Chip Robinson won the GTP race with his Nissan NPT91; David Tennyson was first in Camel Lights with his Ferrari-powered Spice. Butch Leitzinger won IS with his 240SX Nissan and Lance Stewart took a controversial Consulier to the Supercars win. Bryan Herta won Barber Saab and Jovy Marcelo, who would go on to win the season championship, won Toyota Atlantic. Jovy, of course, was tragically killed at Indianapolis the following May.

June 8: SCCA New York Region NARRC.

June 29: SCCA Northern New Jersey Region National. Among the winners were David Lapham (SSA Mitsubishi), Peter Tonelli (SSC Toyota MR-2), Mike Brent (F440 KBS), Steve Johnson (S2000 JRE), Michael Gnazzo (Shelby Can-Am), Paul Newman (GT1 Camaro), Bobby Brown (FA Swift DB-4) and Dave Arnao (GT2 Porsche 914).

July 3-6: Brian Redman Porsche Invitational. Vintage Porsches and vintage Porsche racers came back for a long weekend. Among the attendees were endurance racing stars Brian Redman and Derek Bell.

July 20: SCCA New England Region NARRC. Tom Natasi set a new lap record of 1:03.736 in his E-1 Camaro and Vincent Chiocchi a new record at 1:08.252 in his ITC Fiesta.

August 17: SCCA Mo-Hud/Northern New Jersey Regions NARRC. Paul Roane, Jr. set a record at 1:06.390 in his ITB Volvo, Alan Tackman a record at 1:05.211 in his ITS Datsun 240Z, Wilson Wright a 1:05.494 with his ITA Mazda RX-2 and Phil Phillips a 1:05.810 with his ITE Camaro.

August 30-September 2: BMW Vintage Festival and American Specials Reunion.

September 21: SCCA New York Region NARRC Runoffs. Among the NARRC Champions were Slim Bryant (Spec Racer), Bruce Simon (FP), Dick Clark (GT4), William Grasso (EP), William Collyer (GP), Harold Flescher (HP), Kurt Weiss (ITA), Gary Bossert (ITC), Peter Andrighetti (GT1), Frank Jomosz (GT2), Paul Mantaban (GT3), Bill Larsen (FA), Brian Cunningham (FC), Ed Breault (S2), Jeff Menachem (SCA), Ron Pellman (CRS), Richard Taylor (FV), Bart Golankiewicz (F440), Norman McLeod (ITS), Rick Pocock (ITB), Tom Grossart (FF), David Porter (NCF), Peter Rose (SSGT), Phil Phillips (E1), Gary Attanasio (SSA), Donald Yeske (SSB) and Steve Potter (SSC).

September 27-28: IMSA GTO/GTU. Race winners were Steve Millen (GTO Nissan), John Fergus (GTU Dodge Daytona), Dick Greer (AAC Camaro) and Tony Kester (Olds Pro 2000 Lola).

(Previous page): Second generation racer P.J. Jones, son of legendary Parnelli Jones (winner of the 1970 Lime Rock Trans-Am, among many other races), at speed in his 1992 Toyota MKIII. His teammate, second generation racer Juan Manuel Fangio II, nephew of five-time World Champion Juan Manuel Fangio, won the 1992 Lime Rock IMSA Toyota Trucks Camel GT in his matching Toyota #99. The Toyota team is run by another legend, Dan Gurney.

working every weekend on the race car he drives himself, will become increasingly rare in the future. He just won't command the necessary time nor skills. One-Design racing and Turn-Key racing are already the most important trends in both amateur and professional racing. By the end of this decade, they'll be the norm, not the exception.

Racing In The Future

Lime Rock President Skip Barber spoke to a group of racing enthusiasts recently. His topic? "Lime Rock Park: The *Next* 35 Years of Racing." Among the things he had to say:

"There are a lot of people who think that the height of road racing in the United States was in the Sixties, when the Indy drivers finally ran against the amateurs ran against the professionals ran against the Formula I drivers. That was a Big Deal.

"When the factories all ran against each other in the Trans-Am. That was a Big Deal. When the international driving stars ran in the Can-Am. That was a Big Deal. Formula 5000 was probably the best racing, but we just couldn't sell any tickets. So that wasn't such a Big Deal, even though it should have been.

"Today, twenty years later, things are very different. Any race track needs diversity, it needs both professional and amateur racing. You can't sell tickets and signage and hospitality tents without the big crowds, and you can't get the big crowds without professional racing. That is what the sponsors pay for.

"But we also have to have amateur racing. Out of amateur racing come the professionals, both drivers and mechanics. Out of amateur racing come not just the corner workers and organizers, but even the marketing people who end up at professional teams.

"I'm concerned about the future of amateur racing. The drivers are getting older, the spectators are getting older. Clearly, a major problem for

(Continued on page 200)

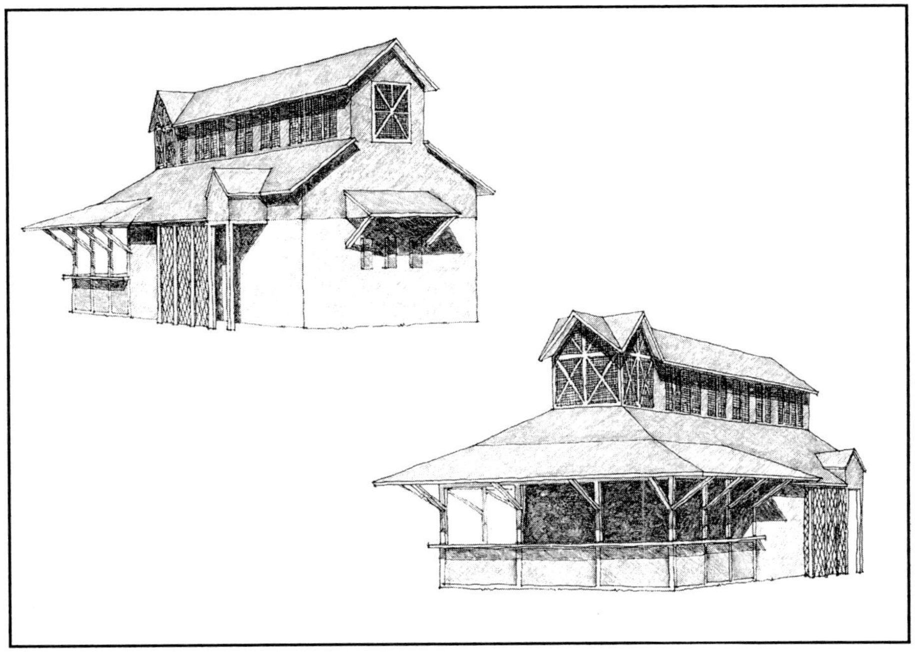

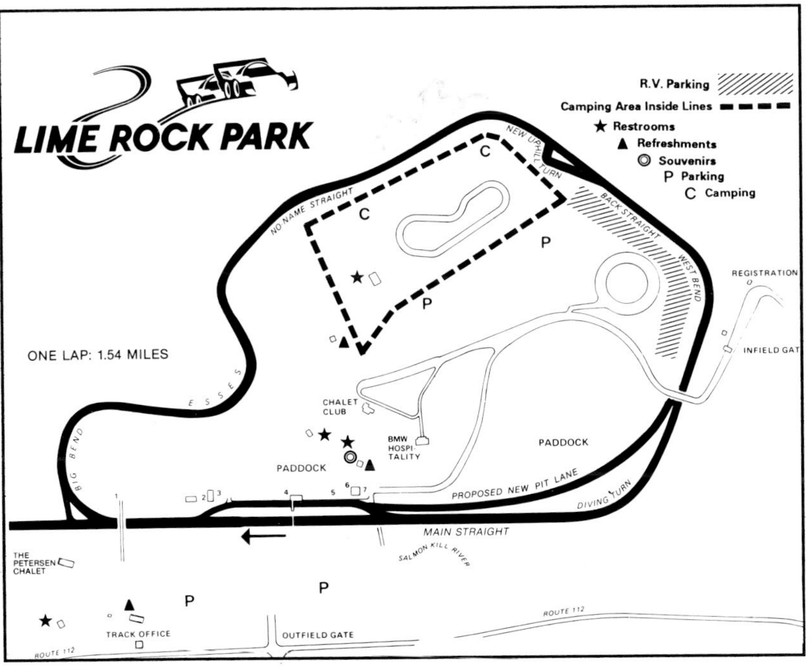

(Left) Plans are to build a new Pit Lane that will exit the course before The Downhill, come through the present False Grid area and connect with the existing Pits. This will be much safer than the current configuration.
(Above) Sam Posey and his half-brother, New Haven architect David Moore, have donated the designs for most of the new buildings at Lime Rock. This charming David Moore sketch remains just that so far.

1992

May 23-25: Toyota Trucks Lime Rock Grand Prix. Polesitter Davy Jones, driving the Jaguar XJR-14 IMSA GTP car—that's the one powered by a Cosworth Formula I V-8—broke the outright track record in practice with a lap at 43.985. He also broke the race one-lap record at 45.857 seconds on Lap 14. He then crashed in The Downhill on Lap 15. After 2 hours of racing, Juan Manuel Fangio II won the GTP race in his Toyota Eagle MKIII ahead of Geoff Brabham-Chip Robinson in a Nissan NPT 92B and Pete Halsmer in a Mazda RX-792P. Parker Johnstone-Dan Marvin were first in the Lights class in their Spice/Acura ahead of Jim Downing-Tim McAdam in a Kudzu/Buick. In the Toyota Atlantic Series race, Chris Smith started from the pole and led every lap. Steve Cameron and Bert Hart were second and third behind him. In the Bridgestone Potenza Supercar race, Shawn Hendricks in a Corvette beat Jeff Purner in a Porsche 911 Turbo and Doc Bundy in a Lotus Esprit X180R. Ashton Lewis, Robert Amren and Alex Padilla were one-two-three in the Zerex Saab race.

How fast are today's IMSA cars? Well, the next weekend, Davy Jones went to Mid-Ohio with a new XJR-14 and not only broke the IMSA GTP lap record by nearly 3 seconds, but also the Mid-Ohio outright track record by .186. The track record he broke was IndyCar National Champion Michael Andretti's, set in a Lola T-91/Chevy A. You can make a pretty good argument, in other words, that the IMSA GTP cars that run at Lime Rock are the highest-performance racing cars in this country, perhaps in the world. Yes, Lime Rock Park *is* world-class.

Class lap records at Lime Rock Park, as of July 19, 1992 (Compiled by Greg Rickes and Craig Robertson):

AP: Panzarella, Corvette, 58.600 (7/5/76)*
BP: Dus, Corvette, 58.340 (9/1/80)*
CP: Slagle, Triumph TR-8, 58.340 (9/1/80)*
DP: Poduje, Lotus Elan, 59.920 (9/6/82)*
EP: Hinkle, Lotus 7, 1:00.110 (5/7/88)
FP: Newcombe, Spitfire, 1:00.200 (7/1/77)
GP: Boig, Fiat X1/9, 1:01.190 (8/30/86)
HP: Wentzell, Sprite, 1:01.310 (8/30/86)
GT1: Bell, Camaro, 53.387 (5/5/90)
GT2: Arnao, Porsche 914, 57.307 (5/5/90)
GT3: Schmid, RX-7, 59.080 (7/5/86)
GT4: Wiley, Datsun 510, 1:01.050 (7/4/87)
GT5: Zimmer, Alfa Romeo, 1:02.687 (5/2/92)
SSGT: Wall, Camaro, 1:02.929 (6/27/92)
SSA: Attanasio, Mitsubishi, 1:06.487 (9/21/91)
SSB: Payson, Mitsubishi, 1:07.141 (7/18/92)
SSC: Tonelli, Toyota MR-2, 1:07.591 (5/4/91)
SSS: Wright, VW Rabbit, 1:13.400 (8/2/75)*
SSSC: Honeywell, Porsche 914, 1:12.800 (7/5/75)*
AS: Cooper, Camaro, 1:03.129 (6/6/92)
IT Enduro: Phillips, Camaro, 1:05.810 (8/17/91)
ITS: Wright, Mazda RX-7, 1:03.820 (9/28/91)
ITA: Weiss, Mazda RX-3, 1:05.280 (9/28/91)
ITB: Pocock, Alfa Romeo GTV, 1:05.184 (9/21/91)
ITC: Bossert, BMW 1600, 1:07.754 (9/21/91)
ASR: Johnson, Lola-Olds, 53.185 (6/8/91)
BSR: Cicale, Chevron, 53.800 (7/5/75)*
CSR: Thompson, Ralt RT-5, 51.580 (7/4/88)
DSR: Miller, Ocelot, 56.850 (7/5/80)
S2000: Johnson, Lola SR71, 52.952 (6/4/91)

SRenault: Lagasse, Sports Renault, 1:01.390 (8/30/86)
Olds Pro: Kroger, Swift-Olds, 51.885 (9/29/91)
Shelby Can-Am: Gnazzo, Shelby-Dodge, 54.083 (6/29/91)
FA: Shelton, Swift DB-4, 47.129 (8/6/88)
FB: Sposato, March, 50.8 (7/7/77)*
F5000: McRae, McRae, 50.000 (9/21/72)*
FSV: Gleason, Modus, 55.000 (9/21/76)*
FC: Rettenmeier, Ralt RT-5, 51.070 (9/2/85)
FF: Rubio, Swift DB-1, 53.830 (9/3/84)
Club Ford: Buck, Lola, 54.449 (8/19/89)
FV: Seibenaler, Zink, 1:00.440 (7/1/89)
F440: Brent, KBS Mark 5, 58.681 (6/29/91)
A-Sedan: Posey, Mustang, 59.200 (5/30/69)*
B-Sedan: Forbes-Robinson, Datsun 510, 1:01.700 (9/4/76)*
C-Sedan: Henderson, Datsun, 1:03.600 (4/5/76)*
D-Sedan: Gross, Mini Cooper, 1:07.400 (no date)*
CM: Barber, McLaren-Chevy, 56.300 (no date)*
DM: O'Brien, Ferrari Dino, 1:02.000 (no date)*
EM: Locke, Porsche 906, 1:02.900 (no date)*
FM: Sachs, Brabham BT8, 1:01.8 (no date)*
GM: Gibson, Lotus 23, 1:03.8 (no date)*
HM: Igleheart, Bobsy, 1:05.400 (no date)*
Can-Am: Jacques Villeneuve, GR-3, 46.930 (7/4/83)*
Can-Am, Under 2.0-liter: Greenfield, Ralt, 48.814 (8/4/84)*
Trans-Am: Dallenbach, Jr., Camaro, 52.760 (no date)
Pro Rabbit: Hacker, VW Rabbit, 1:08.140 (no date)*
VW Golf: Oag, VW Golf, 1:08.777 (no date)*
Racetruck: Krosnoff, Nissan, 1:08.415 (8/6/89)
Canadian Honda: Hebert, Honda Civic, 1:07.420 (5/6/90)
E1: Nastasi, Camaro, 1:03.736 (7/20/91)
Barber Saab: Cochran, Barber Saab, 52.170 (5/30/88)
Firehawk GS: Heinricy, Camaro, 1:04.010 (9/5/87)
Firehawk Sport: Strong, Starion, 1:06.730 (9/7/87)
Firehawk Touring: Green, Honda CRX, 1:07.430 (9/7/87)
Firehawk Compact: Hacker, VW Golf, 1:10.870 (7/4/85)*
GTP: Brabham, Nissan GTP ZX, 46.470 (5/30/88)
Lights: Coppelli, Alba/Ferrari, 51.160 (5/30/88)
GTX: Fitzpatrick, Porsche 935, 49.540 (5/31/82)*
GTS (GTO): Millen, Nissan 300ZX, 50.380 (9/28/91)
GTU: Schroeder, Dodge Daytona, 52.634 (9/21/88)
IS: Shaw, Mazda 323, 1:01.250 (5/25/88)
AC: Danielson, Buick Somerset, 55.340 (no date)
ProFormance: Peterson, Acura, 1:02.080 (5/25/87)*
ProStock: Pobst, VW Golf, 1:10.250 (5/30/86)*
Renault Cup: Knowles, Renault LeCar, 1:12.320* (no date)
Renault Cup: Travis, Renault Encore, 1:11.370* (no date)

1.54-Mile Course with Chicane:
GTP: Jones, Jaguar XJR14, 45.857 (5/25/92)
Lights: Pace, Spice/Buick, 51.260 (5/27/91)
IS: Johnstone, Acura Integra, 1:03.780 (5/26/90)
F/Atlantic: Till, Swift/Toyota, 50.700 (5/28/90)
Barber Saab: Amren, Barber Saab, 54.910 (5/27/91)
Supercar: Bundy, Lotus Esprit, 1:03:515 (5/23/92)

Official One Lap Track Record (set on 1.54-mile course):
Davy Jones, Jaguar XJR14, 43.985, 126.04 mph (5/23/92)

**Indicates inactive class.*

everyone involved is getting younger people interested in amateur racing.

"What's going to happen to professional racing? NASCAR is strong, because they have established their drivers as stars, they have controlled costs and they have wonderful racing.

"There will always be a major single-seat racing series in this country, whether it's called CART or IndyCar or something else. As long as there is the Indianapolis Motor Speedway, there will be some sort of Indy racing.

"A category of racing that will always be important to Lime Rock is Prototype or Sports/racing, whether it's IMSA or Can-Am or something else. I think that what will happen there is that there will be a Prototype class where a privateer can buy a chassis, buy an engine and go racing. Sort of a NASCAR approach.

"The other category that I think we'll surely have in the future is some sort of road racing for sedans, whether it's Trans-Am, GTO or whatever.

"A big problem with all our series is that there is no road racing sanctioning body capable of turning a driver into a star. Just about every car manufacturer in the world is in involved in road racing in one way or another, but the stars are not there to draw the crowds."

The unchanging face of amateur racing; Michael Caron checks the schedule before taking his Spec Racer out for a NARRC Regional, 1992. Yes, it's still possible to go racing with a van, a trailer, a few spares in milk crates, not much money but a lot of enthusiasm.

Lime Rock In The Future

"I feel much better about Lime Rock Park's future than I do about the overall situation. Lime Rock has some significant strengths that allow us to survive bad times and prosper in good times. The most important is that Lime Rock is so short. That makes it a great place for spectators because they can sit on the Hill and see half the course at one time, and they can see the cars forty times in a half-hour race. That's neat. Lime Rock is also a pretty place to go; a place you can bring your wife and family for the day.

"Because it is small, Lime Rock costs less to maintain than someplace like Road America that is 4 miles long. We need fewer guards on race day, it costs less to keep the

(Continued on page 202)

FOR THE RECORD

Holders of the outright track record at Lime Rock Park. All records were set during a race.

1957: Walt Hansgen, 1:10.40
1957: Carroll Shelby, 1:06.00
1958: George Constantine, 1:05.81
1959: George Constantine, 1:04.83
1961: Roger Penske, 1:04.30
1962: Roger Penske, 1:03.40
1963: Jim Haynes, 1:02.90
1963: George Alderman, 1:02.00
1964: Peter Sachs, 1:01.80
1965: George Alderman, 1:01.60
1966: Don Morin, 1:01.50
1967: Sam Posey, 58.0
1967: Skip Barber, 56.30
1968: Lou Sell, 54.80
1969: Peter Gethin, 52.90
1969: Andrea de Adamich, 52.10
1970: David Hobbs, 50.80
1972: Graham McCrae, 50.00
1978: Howdy Holmes, 49.09
1980: Jeff Wood, 49.04
1983: Jacques Villeneuve, 46.930
1988: Geoff Brabham, 46.470
1992: Davy Jones, 45.857
1992: Davy Jones, 43.985, Quickest Single Lap. (Jones' records were set on the 1.54-mile "long" course, which adds about 2 seconds to lap times.)

(Far left) Juan Manuel Fangio II.
(Below) The Toyota Eagle MKIII with which Fangio II won the Lime Rock Park IMSA GTP race on Memorial Day, 1992.
(Left) The Jaguar XJR14 with which Davy Jones set a new qualifying track record of 43.985 on May 23, 1992 and a new race track record of 45.857 on May 25, 1992.

lawns mowed. You can cover a race at Lime Rock with five television cameras. At Road America, you need fifteen. Lime Rock Park has a lot of things going for it that are going to be important in the Nineties.

"In the Nineties, we are going to improve spectator access and parking, we are going to make it so that spectators can walk all the way around the perimeter of the Infield and watch the races. We are trying to buy another piece of adjacent property so that we can have another spectator entrance and a spectator field on the outside of The Downhill.

"The most important thing we will do in the immediate future is repave The Downhill and The Straight. The second most important thing is a new Start/Finish Building. We will do some more paving in the paddock, and construct a new Press Building. We are working on maintaining good relations with the town and the state, and improving relations with our neighbors.

"We are experimenting with different dates for professional races, trying to ensure that we have more than one big weekend. Right now, we have the IMSA GT weekend on Memorial Day and the BMW Vintage Festival on Labor Day. We would like to have at least one more major event. Are we going to be a two-race track in the future, a three-race track, a four-race track? I don't know. We're still experimenting."

A Little Jewel

"I think of Lime Rock as a little jewel. In the future, we're going to make it *so* nice, make it such a nice place to spend the weekend that families will go out of their way to come here. There are things to change. We have to give the racers a smooth track and a nice paddock area, we have to give the spectators a pleasant place to visit, we have to give them all superlative food, we have to give them exciting racing. That's the future of Lime Rock Park. It looks good to me."

9. Vintage Racing

Even vintage racing has its history. This Model T Speedster was photographed racing down No Name Straight in 1965 or so. We don't know anything about the driver, his comely passenger or the car itself except that, together, they've got an almost unbearable amount of charm.

(Top) George Alderman and his Capri-panted pit crew, after winning the Formula III race at Lime Rock in 1959 with their Norton Manx-powered Cooper.
(Bottom) The very same Cooper in a VSCCA race at Lime Rock 25 years later. Yes, you can recreate history!

Anyone who was there can shut their eyes and bring it all back; the distinctive symphony of the engines—basso profundo American V-8s, shrill trumpets of Ferraris, lovely, high-pitched woodwinds called Coventry-Climax—the raw stink of hot Castrol R, the sweet crowd smell of Hawaiian Tropic and perspiration, the blur of primary colors as the cars flashed down the hill and onto the Main Straight. You could stand by the side of the track and watch forgotten heros like Lance Reventlow and Pedro Rodriguez sawing away at the wheel, cornering in a series of tire-twisting skids, your heart in your mouth along with them.

Despite the thrills and danger, it was a kinder, gentler time. A time when racing was more of a sport than a business, when drivers could be fat and tires were skinny and you could drive home in the car you raced, carrying your entire pit crew with you in the other seat, her hair kept neat beneath a broad-brimmed hat.

Vintage racing is about bringing all those memories and experiences back to life again. "You can't relive the past," explains vintage Porsche racer Frank Gallogly, "but you can sure *recreate* it." Particularly at Lime Rock Park, which has changed so little over the past four decades, it is possible to select a spot and snap a photo—mental, Kodak or Sony—that is literally unchanged since the same car went through the same corner thirty-five years ago. To a certain type of car enthusiast, such moments are more precious than rubies.

The Boom in Vintage Racing

Vintage racing is the fastest growing area of motorsport. Irreplaceable old machines that by right have earned a quiet rest are being dug out of barns, lavishly restored and trailered to the very tracks they snarled around decades ago. A vintage racing cars wants to be driven, *needs* to be driven. None were ever designed to be sequestered in a museum.

The key to keeping vintage racing authentic is that, unlike even amateur racing these days, it is purely for fun. There is no "championship," no purse; often there's not even a trophy to tote home. Cars sport numbers not cigarette logos. The point is not necessarily to win but to show off priceless racing machines the way they were designed to be seen—sunlight smiting the chrome, exhausts shrieking and over all, the aroma of hot Dunlops and thoroughly baked Ferodo DS-11.

The whole atmosphere around vintage racing is different, and it's an

(Continued on page 209)

(Far left) The Concours d'Elegance at the 1991 BMW Vintage Festival stretches the entire length of the Main Straight.
(Left) Don Lefferts' Bugatti Type 35. This car has been racing at Lime Rock since the place was built.
(Below) David Brownell, editor of Hemmings Motor News *and organizer of the flea market, at the 1991 BMW Vintage Festival.*
(Bottom) Bob Girvin in his rare Allard coupe heading out for practice, 1991.

58

racing weekend on the East Coast.

What can you say about a race weekend that stretches over four days, attracts over 300 priceless vintage and historic race cars from literally all over the world and plays to an audience of tens of thousands of spectators and millions of TV viewers worldwide. Every car accepted for Vintage Festival races at least twice, plus practice and qualifying sessions spread over Friday, Saturday and Monday.

(Previous page) VSCCA Director Charlie Gibson at speed in his potent Lister-Chevy...the very definition of vintage racing.
(Above) The ultimate vintage racer, Porsche 917.
(Right) Three Cunningham C4Rs—including the Kamm-designed C4RK coupe—lead the Michelotti-designed C5R into the Lime Rock pits at BMW Vintage Festival, 1991. This was the first time Phil Walters had been behind the wheel on a race track since 1955!

MALCOLM LABATT-SIMON

Vintage racing cars. Vintage racing drivers. Why not a vintage racing starter? Malcolm Labatt-Simon is as much an institution at Lime Rock as The Downhill or Big Bend. Every racing driver recognizes Malcolm...he's the one on the Lime Rock starter's bridge giving you thumbs up after your race, even if you've made a total balls-up of the past half-hour. How does one get to do that? What's the career path for a Chief Starter?

Well, probably not what you'd expect. Malcolm was racing Formula Vee in the early-Sixties—he even won the Northeast Division Championship one year—but as he says, "racing got a little bit too expensive at the time. I was Communications Manager of WQXR, the classic music station of *The New York Times*. It sounded great, but it didn't pay too well.

"So I got involved in flagging. I became Chief of Flagging and Communications for the New York Region. In 1966, I won an SCCA National Award for pulling a driver out of a burning car at Station 8 at Lime Rock. I was burned, not seriously. Unfortunately, the driver later died."

Malcolm then became one of the starters in the New York Region. Soon after, Chief Starter Don Brody quit, and Malcolm was Chief Starter. Pretty soon, the SCCA's Northern New Jersey Region called and Malcolm became Chief Starter for them, too. Then he started doing New England Region races, too. Then the VSCCA called.

"The next thing I knew," he says, "I was busy every weekend. It cost me more to go to all these races as the Chief Starter than it cost me to go racing myself. But by then, I was committed." Today, Malcolm Labatt-Simon is easily the best-known Chief Starter in the country. At a youthful 72, he's also been doing it longer than anybody else, too.

At the same time that he was spending every weekend between March and November at the races, Malcolm was starting his own advertising agency, Speigel-Labatt-Simon. Now he has his own small agency, among whose clients is Hertz Equipment Rentals. Two decades ago, he also became a local resident.

"We were at the party after Jim Haynes married Toni," he says, "and Jimmy said 'Have I got a house for you.' My wife Barbara fell in love with it, and we snapped it up. Our residence is at the end of The Downhill, just across the Salmon Kill. Years ago, when Brian Redman crashed going down the hill, he ended up on our property. We told him we had salvage rights and were going to claim the car! When he was younger, my son wanted to put up a toll booth at the end of The Downhill, but we persuaded him not to." ■

(Above) Colorful Malcolm Labatt-Simon has been starting—and finishing—races at Lime Rock for three decades. His traditional "thumbs up" is a welcome sign as you come under the starter's bridge, the occasional "thumbs down" an order to reflect on your sins.

(Right) VSCCA president Gil Steward at speed in his fabulous Packard Speedster, October, 1991. (Below) Pace Lap at the PaineWebber Invitational, Memorial Day, 1991.

There are handsome trophies and at least in 1991, massive jerobaums of Moet & Chandon, given not to the winners of the big races on Monday, but to the drivers who best exemplify the ideals of vintage racing in the way they present both their cars and themselves.

In addition to the racing, there is a huge automotive flea market that goes all weekend, an exhibition of automotive art and on Sunday, an extravangant Concours d'Elegance that stretches the length of the Main Straight. Always there are dinners and parties, with featured guests ranging from Carroll Shelby to Rodger Ward to "Pappy" Hough.

And there is always a special "theme." In 1990, for example, Lime Rock's Mike Joy recreated as closely as possible the 1970 Trans-Am race as a twentieth anniversary celebration. The year before, on the thirtieth annniversary, Jim Shane and I organized a recreation of the famous 1959 Formula Libre race with nearly every car and every living driver who was in the original event.

We even had the original track announcer, Martin Burgess, who flew in from England for the occasion and such subtle touches as the late George Constantine's mechanic Rex Woodgate driving the famous Aston Martin DBR-2. Not only did we have winner Rodger Ward and Kenny Brenn's midget that he drove that day so long ago, but car owner Kenny Brenn, too. If the 1959 Formula Libre was the best race ever held at Lime Rock, the 1989 Formula Libre recreation was surely the most moving.

Putting Dick Thompson, Paul O'Shea, Gordon MacKenzie, John Fitch, Denise McCluggage, George Waltman and Rodger Ward back on the track in the cars they had driven three decades before far surpassed any "Senior Tour" in stick and ball sports for sheer nostalgia and good times. We even got the famous wild man Boris Said to come back and stand around, just like he did in 1959 after his car broke in practice. You can't produce a more authentic recreation than that! The 1989 BMW Vintage Festival Formula Libre recreation was the very definition of what vintage racing is supposed to be all about. When starter Malcolm Labatt-Simon threw the green flag, there wasn't a dry eye in the house.

Vintage Racing Center of the East

Lime Rock Park has become the vintage racing center of the East Coast for a variety of reasons. For one thing, Lime Rock is an authentic track from the vintage racing period, about the only one still around in pretty much its original configuration. It even looks pretty much the way it did back then, or at least close enough to fool the mind's eye.

Even better, as vintage racer Alex Dearborn points out, Lime Rock is the only track where his wife will come to the races! She's not the only racer's wife who feels that way.

"As far as I'm concerned," says Alex, "the best thing that ever happened to Lime Rock was when they banned Sunday racing. Susan and I can drive

(Continued on page 218)

(Left) One of the benefits of writing a book...you get to put a picture of your own car in it! Author Rich Taylor in his 1967 Corvette, on his way to winning The Goodyear Cup at BMW Vintage Festival, 1991. (Below) Taylor again, this time in his 1958 Devin SS, participating in the parade of American Specials at the same event. His passenger is Jean Harker, patroness of the annual Paul Pettey Award, given for the best performance by a Jaguar at BMW Vintage Festival.

(Bottom) Formula Juniors and a Bugatti together in the pits; it must be the VSCCA! (Right) Astonishingly fast, astonishingly beautiful, Bugatti Type 35 in The Downhill.

BMW VINTAGE FESTIVAL

Jim Haynes came up with the idea to hold a purely-vintage racing weekend in 1982, combined with the Lime Rock Concours. Sanctioned by the Vintage Sports Car Club of America and run by Charlie Gibson, the Vintage Fall Festival picked up sponsorship from BMW in 1986. After that, there was no stopping it. BMW Vintage Festival is now the second most popular event at Lime Rock after the IMSA Camel GT, and by far the fastest-growing. Around 300 vintage race cars show up, plus there is the Sunday Concours, an automotive art show and a four-day automotive flea market. Tens of thousands of spectators come for the show.

Starting in 1986, the organizers also started presenting not only awards to the first three finishers in each race, but a silver cup to "the driver most exemplying the vintage spirit, not necessarily the winner of the race." Today, these Cups have grown both in number and prestige. Depending on the year, each is also accompanied by a bottle of champagne. The most prestigious is the Jim Patterson Award, given by BMW.

BMW Vintage Festival
Major Award Winners

1986
Almaden Cup: Bob Girvin
BMW Cup: Sid Herman
Kendall Cup: Robbie Robinson
Lime Rock Cup: Jim Donick
RJ Reynolds Cup: Cap Chenoweth
Patterson Award: Robbie Robinson

1987
Almaden Cup: George Bruggentheis
BMW Cup: Jim Freeman
BF Goodrich Cup: Marv Rosenberg
Kendall Cup: Kent Bain
Lime Rock Cup: John Schieffelin
Lowenbrau Cup: Tony Podell
Patterson Award: Bob Colaizzi

1988
Kendall Cup: Oliver Clubine
Bosch Cup: Bruce Jennings
Miller High Life Cup: Don Nattrass
Lime Rock Cup: Jim Bok
BMW Cup: Mike Rand
Goodyear Cup: Marc Perlman
RJ Reynolds Cup: Ed Henning
Patterson Award: Bib Stillwell

1989
BMW Cup: Leon Hurd
Bosch Cup: Dave Handy
Goodyear Cup: Keith Goring
Kendall Cup: Tivvy Shenton
Lime Rock Cup: Don Funke
Miller High Life Cup: David Paqua
RJ Reynolds Cup: Bob Millstein
Patterson Award: Joel Finn

1990
BMW Cup: Bill Holmes
BMW Cup: Reg Cunningham
BMW Cup: Jeff Gerken
Goodyear Cup: Watts Hill
Kendall Cup: Phil Lamont
Liquid Glass Cup: Kendall Merritt
Liquid Glass Cup: Dick Merritt
RJ Reynolds Cup: Syd Silverman
Barber School Cup: Bob Hollander
Toyota Trucks Cup: Dan Butt
Zerex Cup: Bob Burt
Patterson Award: Nick Jones

1991
Automobile Cup: Don Mei
BMW Cup: Peter Krause
Goodyear Cup: Rich Taylor
Kendall Cup: Gerry Morici
Moet & Chandon Cup: Frank Gallogly
PaineWebber Cup: Gil Steward
PKG Cup: David Kopf
Barber School Cup: Kendall Merritt
Toyota Trucks Cup: Eric Wensberg
Zerex Cup: Harry Gaunt
Patterson Award: John Harden ■

down from the Boston area on Saturday morning in our race car. I can practice, qualify and race on Saturday. Saturday night, because I don't have to race the next day, we can have a nice dinner with a little wine, stay in a pretty country inn, sleep late on Sunday and then have the day to do something fun before we drive home. It makes a perfect weekend that you just can't have at any other race track."

The VSCCA

Perhaps most importantly, Lime Rock is the home track of the Vintage Sports Car Club of America. The VSCCA has been conducting old car events at Lime Rock for nearly thirty years, and it is the only strictly vintage club in this country with a cutoff at 1959. Other, younger old car racing clubs will let you race yesterday's IMSA car; by VSCCA standards, a car plastered with sponsor decals from a professional racing series is not a vintage racing car. The VSCCA is also the only racing group in the country seriously interested in preserving and racing prewar automobiles.

The Cars, The Races, The People

Most vintage racers will tell you they originally got interested because of the wonderful old cars, then they found out that going fast and competing on a race track is a terrific adrenalin rush. But the reason they keep coming back year after year is for the people they've met through racing. With only a few exceptions that prove the rule, some of the nicest and most interesting and thoughtful people you'd ever want to meet are the ones involved in vintage racing. There is a real feeling of camaraderie among most vintage racers.

Obviously, just about anybody who gets near a race track, from the greenest corner worker to the old pro IMSA star, is going to be an interesting person. It kind of goes with the territory. They're extraordinarily bright, dedicated and hard-working, most of them. But vintage racers seem to add another dimension to the usual racing mentality. Racers are constantly looking ahead and obsessed with detail, so they often can't see the longer perspective that assures them a place in history.

Vintage racers, by definition, are not only looking forward to their next race, but looking back thirty-five or fifty or seventy years, too. Not only are they trying to solve the problems inherent to any racing program of any type, they're trying to put themselves into the coveralls of an engineer, driver or mechanic who's been dead for twenty years and figure out how he wanted them to solve today's problem. There are important lessons to be learned from these forgotten racers of the past, for racing history does in fact travel in a direction, like a beam of light.

Character

Ultimately, it's character that defines a vintage car. It is something to believe in, something you can get into—literally and figuratively—and go somewhere. Unlike modern racing cars, which are often indistinquishable even to experts, vintage cars are distinctive, loud and exciting. You can lift the hood and comprehend what goes on in there. And even better, you can usually fix it yourself.

Out on the track, vintage racers slide around in a way that modern racing cars would be embarrassed to do. But all that merely brings an even wider grin to the driver's face as he once again careens into a corner, side exhausts blaring, the hot wind in his face smelling of youth and joy and a better time and place. Vintage racing can bring back dreams of glory, relived in the cars of one's youth. For that alone, it's cheap at any price.

10. One Lap of Lime Rock

Author Rich Taylor in the Lime Rock Park pits before starting a Skip Barber School Formula Ford race in 1983.

0.0
109 MPH

Skip Barber School Senior Instructor, Bruce MacInnes, takes us for One Lap of Lime Rock. Section times and speeds are for a Skip Barber School Formula Ford, 105 hp, street radial tires. Compiled by Carl Lopez, Skip Barber School Director of Planning.

Dry: We're looking at the start/finish line on the Main Straight, just past the bump from the access road that goes across to the pits. With a ground-effects car, you try to drive around and avoid that bump because it upsets the car. In most race cars, you're still accelerating at this point, but near the fastest speed you'll hit on the Lime Rock course.

Rain: Lime Rock has been around a long time, and the normal line is polished. When it's wet, you're better off where the surface is granular.

2.0
110 MPH

Dry: Now we're approaching Turn One, further down the Main Straight. The track actually gets narrower around the "3" marker, so you have to be careful not to drop your left wheels off the edge of the pavement. Our whole philosophy about Turn One has changed since we've gotten our computer-instrumented race car. In the Skip Barber School, we teach a skill called "Trail Braking." The computer shows us that this is what we call a "Type Two" corner where you want to have a high entry speed, but you don't care so much about your exit speed. This means we want to carry our velocity down the Main Straight as long as possible before braking for Turn One. Our whole purpose in racing is to extend our time on the straights and minimize our time in the corners.

2.6
111 MPH

Dry: We brake as late as possible to extend the straight as much as possible, then carry our braking as we go under the Camel Bridge. In a Formula Ford, we start braking between the "1" and "2" markers. We're very careful to release the brakes smoothly, because the release of the brakes creates oversteer. This creates "rotation," meaning the car will rotate on its axis. We like to have a little bit of rotation; that helps position the car for the corner. Too much rotation, of course, and you'll spin. In a Skip Barber Formula Ford, we simultaneously shift from Fourth to Third; in a National-level Formula Ford, you'd go from Fourth to Second. We carry a lot of brake/turning into the corner. We actually start turning before the "1" marker. There's a big white blotch marking the spot.

5.5
72 MPH

Dry: We treat Big Bend as two separate corners. We clip the first apex where the cone is, then let the car come out a little bit. You don't want to get more than two-thirds of the way to the outside of the track; there are always a lot of "marbles" out there. We try to use a little bit of trailing throttle to put weight onto the nose for better bite. Some production car drivers will even left-foot brake to accomplish the same thing.

Wet: We stay inside the normal braking zone coming into Turn One, then cross the polished pavement to the outside. We call this "the slide for life." You end up almost on the apron for the little connector road on the outside of Big Bend. Then you run around the outside of the corner all the way to Turn Two.

6.5
67 MPH

Dry: We stop trailing the brakes just before the first apex and squeeze on the power. If you can go back to full throttle you've gone in too slow; if you can't add any power, you've gone in too fast. It's hard to believe, but the second part of Big Bend is actually banked and uphill. You can get in low and early and get away with it. Almost all of Big Bend is a passing zone. You can go inside approaching the first apex and pass people under braking. You can go by them on the inside through the turn right up to the second apex. People do it, but passing on the outside of Big Bend is just asking for an invitation to visit the grass. Of course, in the rain, the fast line is around the outside, so you may very well pass people while you're out there. Just hope they don't move out on you.

11.0
65 MPH

Dry: Now we're headed towards the second apex of Big Bend, also called The Hook. If you think of this as one big corner, the second half has a decreasing radius. That's why we split it into two parts. You can see that the yellow curb has a rounded part and then a flat section. The safe bet is to apex the flat section. About halfway out of the turn, there's a seam in the pavement and then the track flattens out. To be safe, you want to stay inside this seam. If you get outside it, you may slide to the outside.

Wet: A lot of guys crash in Turn Two when it's raining. They get power oversteer, try to hold on and crash, Big Time. Let it go wide and run it in the dirt, if you have to! That's better than augering into the guardrail trying to stay on the pavement.

14.0
68 MPH

Dry: Probably the most common incident at Lime Rock is that racers get in too early for the second apex of Big Bend, then run wide coming out and drop their two left wheels off into the dirt about where the cone is in this picture. Then they lift off the throttle. That plants the nose and they hook-slide across the track into the inside guard rail, often collecting somebody else in the process. Ouch! If you drop two wheels off, the proper thing to do is steer a little to the outside and go to even throttle. If you get more than say 40 degrees out of shape, just stand on the brakes and lock 'em up. You'll spin harmlessly to the outside of the track on the grass. When the grass is wet, you may actually accelerate when you hit it, so this is not such a good plan.

16.5
73 MPH

Dry: You can use every inch of the short straight between The Hook and The Left-Hander, cutting diagonally across it from left to right. We aim for the white door on the side of the Driving Impressions building. The idea is to keep to the outside and avoid an early apex. If you consistently early apex, brake later and you'll be *forced* to turn later.

Wet: There are two philosophies about The Left-Hander in the rain. One says that in our Formula Fords, we should stay below the seam. Approaching the Left-Hander, we never get above the seam. We can even double-apex this corner. You go into it early, brake and turn, go across the apex—which is almost the worst place on the track in the rain—and do "the slide for life."

18.0
51 MPH

Dry: You can clearly see the seam in the middle of the track in this picture. A good rule is to stay outside that seam until you go past the white patch in the track. In a fast car, The Left-Hander can be a "Type Three" turn that you "throw away" in order to get positioned for Turn Four, which is much more important.

Wet: Once you slide across the apex, you pick up the granular surface on the outside again and find your rainy day traction. We didn't know any of this before we got our onboard computers. I did "segment times" in the rain, and my worst student in the dry had the best times in the rain. He didn't know where to drive, so he was driving off the line where the traction was better in the wet. Amazing. He was .6 seconds faster from Turn Two to Turn Four.

19.2
55 MPH

Dry: We're right on the apex of The Left-Hander. Perfect. Let the car come out about one-third of the way to the right, no more. Turn Three is very bumpy, with a lot of laddering from the GTP cars. With our Formula Fords, we try to run through it or a little bit inside it, though in a big car you might have to straddle it. In our Skip Barber Formula Fords, we stay in Third gear all the way from Turn One through Turn Four. In a National Formula Ford, you might gear it to use Second gear instead.

Wet: After we do "the slide for life" across the apex of The Left-Hander, we stay to the right on the Short Chute. We go back on the power, hustle across the normal line and approach Turn Four on the left, outside the normal line. It may seem strange, but it works.

21.0
57 MPH

Dry: Turn Four, The Right-Hander, is a "Type One" corner that leads on to a long straight. At Lime Rock, the second half of Big Bend, Turn Four, The Uphill and The Downhill are the only Type One corners. So that makes them the key spots on the track. It's very important that we position ourselves perfectly into Turn Four, which can be very difficult. The idea is to sacrifice entrance speed if necessary in order to gain a higher exit speed coming on to No Name Straight. Coming out of Turn Three, we stay to the left third of the track, even though that slows us down at this point.

Wet: Lime Rock is a track that rewards smoothness at any time, but particularly in the rain. In the dry, you can get away with being rough through The Esses. Not when it's wet.

22.0
61 MPH

Dry: We stay to the left of the seam until very late, then turn in to hit an apex in the second half of the yellow curb at the inside of Turn Four. On a sunny day in any season, there is always glare through here that disorients some drivers. It gets worse later in the day when the sun can be shining right in your eyes. This is a good place to pass on the inside; instead of sacrificing speed through The Left-Hander, you can keep your speed up on the Short Chute, duck inside in The Right-Hander and seize the line coming on to No Name Straight. The trick is not to come out of The Right-Hander slowly that you get passed again on the inside going down the straight. This can be one of the trickiest spots at Lime Rock on the first lap, because the cars tend to bunch up through The Esses.

23.8
65 MPH

Dry: Right on the apex of Turn Four. Perfect. The track is rough through here, so you want to stay inside the "laddering" if you can, or straddle it. This is such an important turn, because it determines your speed down No Name Straight, into The Uphill and all the way to West Bend. Screw up Turn Four, and you've blown half the lap before you get your speed back again.

Wet: There's a low spot right on the exit of Turn Four that fills with water and becomes a puddle. You have to watch out for it; it will put you right off. The whole concept of driving this track in the rain is to go from one outside to another outside, spending as little time as possible on the dry line. That will let you put the power down. In the rain, the normal line will get as greasy as an ice-skating rink.

25.0
67 MPH

Dry: We let the car go all the way out to the left edge, using all the track. This is another spot, similar to the outside of Turn Two, where a lot of drivers will put two wheels in the dirt, lift off the throttle and hook to the inside of the course. Sometimes they'll take somebody else with them. Before they cut the bank square, a Corvette got off on the left side, hooked to the right and actually went through the wire fence and ended up near the spectator area on top of the infield hill. If you do get off, steer slightly to the left and steady the throttle, then smoothly come back onto the track. Actually, if you leave the steering wheel alone, most cars will usually bring themselves back onto the track, even though you may be steering a little to the left. Whatever you do, don't overcorrect.

26.6
72 MPH

Dry: A great speed trick at Lime Rock is to completely ignore the right-hand apex of The Kink on No Name Straight. Instead of weaving across the straight, hit the left apex at the cone in our photo and then stay in the middle of the track. Our onboard computer shows that it costs about 200 rpm to pitch the car off to that right apex. It will take you to West Bend to get that speed back again.

Wet: We stay way over to the left all the way down No Name Straight. No Name Straight is the edge of the earth in the rain. More people crash here in the wet than anywhere else on the course. It's amazing. They say accidents happen where you least expect them. It's true. In the rain, they happen on straightaways, not in the corners.

28.6
75 MPH

Dry: This telephoto shot shows exactly why we don't want to apex the right hand side of The Kink. See how bumpy it is? That's why we stay towards the middle/left of the track away from this area. This also shows the way No Name Straight pitches from side to side. The first half is cambered to the left, the second half is cambered to the right. A lot of debris also seems to collect here. There is very little run-out room on either side, which makes this one of the most difficult spots at Lime Rock if something goes wrong. Because it is a major passing zone—and a difficult one, too—there seem to be an inordinate number of "incidents" in this area of the track. At sunset, the sun is also directly in your eyes as you come down the straight, which seems to create problems, too.

30.5
80 MPH

Dry: Now we're apexing the left-hand side of The Kink, halfway down No Name Straight. If you look at the dirt bank on the infield side, you'll see what look like erosion marks. I call that "carosion." It's from race cars augering into the bank. This is a much trickier part of the course than most people think...until they've crashed here either in the rain or traffic. In our Skip Barber Fords, we stay in Third gear through here—if you can go fast enough in Third to blow the engine, I'll buy you a Coke.

Wet: No Name Straight is a big problem when it's raining. It is very polished and *very* slippery. You have to stay left, way left, all the way down to the turn-in point. You'll have to cross the normal line; be careful, let it rhythm out, then come back to the turn-in point.

33.3
89 MPH

Dry: The approach to The Uphill is tricky. We want to be as far to the left as possible when we get to the turn-in point where the cone is in this photo. If you drive what feels like a straight line from the inside apex of Turn Four through the outside apex of Turn Four and the left-hand apex of The Kink to the turn-in point of The Uphill, the left-hand edge of No Name Straight will seem to ebb and flow up to the side of your car and you'll be in just the right position for The Uphill. In traffic, you may have to modify your line, otherwise overtaking cars can slip inside you going into The Uphill. This may be the fast line in the rain, but it's an invitation to visit the outside guardrail in the dry. Happily, you can always scrub off speed going into The Uphill if you get into trouble.

33.8
90 MPH

Dry: You either love The Uphill or you hate it. I love it, but it's also the only turn that has ever put me into the hospital. Big Time. I tried to take it flat-out in Fourth gear, the upshift was my "lift." That didn't work. Now I know there are only four possibilities when someone tells you he's going through The Uphill "flat-out." One possibility is that he *thinks* he's flat-out. Another is that he's going so slowly, because he blew The Esses so badly, that he doesn't have to slow down by the time he gets to The Uphill. Another is that he was going flat-out *into* The Uphill. Before he crashed. The fourth is that he lies. If you're going fast enough, you always have to either touch-and-go on the brakes or do an early "feather" of the throttle at the end of No Name Straight.

35.7
84 MPH

Dry: Here we're right at the apex. Excellent. The Uphill is actually *two* separate turns. One turn gets you from the turn-in point to the apex. Now you can see that there are three white lines going across the track. At the first white line, just past the apex, we make a second, sharper turn. We want to have the car parallel to the edge of the track when we cross the second white line.

Wet: There are two philosophies on The Uphill in the rain. One says to straddle the wagon wheel ruts and brake straight to the rim, then drive right across the outside. This is a safe bet. It's great. The other philosophy says to hug the inside, do a lot of braking on the inside, then do "the slide for life" straight toward the guardrail. Not many people prefer the second option except me.

36.7
83 MPH

Dry: We want our race car to be positioned just to the left of the white patch in the track, with our car parallel to the edge of the track, as we go up the hill. In this photo, the car is actually a little bit on too much of an angle across the track. It's better to be parallel to the track in the middle of the hill rather than still turning at the top.

Wet: The outside of the track is pretty bumpy here, but you can still get superior traction in the rain. Right here is where I won the Can-Am in 1985 in a rain storm so bad they stopped the race a few laps later. I only had a little cast iron engine in that car, but I knew how to do the rim shot and I was able to pull past this guy on the outside up the hill and onto the Back Straight. He never saw me coming.

37.2
84 MPH

38.5
86 MPH

40.0
88 MPH

Dry: Now we're okay. We're parallel to the track and near the left edge. We can be going faster through here than you'd think, because the compression effect of the hill is against us. Just over the brow of the hill, there's an apron on the left side. You *don't* want to hit that; the leading edge can break a suspension. Even many experienced drivers slow down too much at the bottom of the hill and never get their speed back until West Bend. If you are fast up the hill, you can often slingshot past slower cars at the top, seize the line and pull away from them into West Bend. In that sense, The Uphill and the Back Straight are one of the best passing zones at Lime Rock. You almost always pass on the right, which puts you on the inside going into West Bend.

Dry: At the top of the hill, your car will get light on the suspension, and if you're not parallel to the edge of the track, it may feel like it takes a step to the left. If you've early-apexed The Uphill and you're still driving across the track, of course, you'll have to make a correction just when you're light, and you can easily spin to the inside. You can see in the photo that the left edge of the Back Straight is not even. If you get too far left, you can easily drop a wheel off here. It's better to leave a margin of a couple of feet and drive straight from the top of the hill to the turn-in point for West Bend than waste time following the edge of the road. Basically, you can aim a little to the right of the Camel GT billboard when you crest the hill. That will put you right when you reach the turn-in.

Dry: The turn-in point for West Bend is marked by a white blotch of paint. A common question is, "How fast do you go into this corner?" Some racers don't brake at all, some brake heavily, some just tap. It really depends on the car and the speed you've carried up the hill. If you're slow up The Uphill, obviously you won't need to slow for West Bend. West Bend is also not as sharp a corner as it appears; it's less than 70 degrees. I'd say most people come through here more slowly than they could go. This is not a particularly good passing area if the cars are pretty equal in speed. People expect to pass under braking, but nobody really brakes. It's safer to pass either on the straight or after West Bend, going under the Toyota Bridge when you can easily cut to the right and pass down the hill.

41.0
88 MPH

Dry: Right here, I pivot the car with the brakes, just a touch-and-go immediately before the turn-in point to help position the car. I want to get the nose down to help me turn. In our Skip Barber Formula Fords, we're still in Third gear at this point. It wastes too much time to shift up and back down again, unless your car is geared so low you're going to blow the motor if you don't shift. Of course, in faster cars like IMSA GTP machines, you'll typically shift up a gear on the straight and hold that gear through West Bend. Once again, it really depends on the car's balance between power, gearing, tire adhesion and many other factors. This is a good place not to lose momentum by going too slowly into West Bend. That will slow you all the way down the hill.

42.5
86 MPH

Dry: Immediately, as soon as we tap the brakes and turn into West Bend, we're back on the power and tracking through the corner, using all the road. We're aiming at the apex, two-thirds of the way down the yellow curb. You don't want to early-apex West Bend; you'll end up out in the tire wall at the outside of the turn. You can see where the tires have been rearranged by drivers who've made just that expensive mistake.

Wet: West Bend is very curious when it's raining. Rather than go off-line, you stay on the normal dry line through the turn-in point and all the way around the outside. Just beyond the turn-in point, the track gets very granular even on the dry line, and you can still get traction when it's wet.

44.0
87 MPH

Dry: Perfect. We're just coming up to the apex. See the dip in the pavement just past the apex? We'll just drive through it, though you might want to avoid it with a ground-effects or production car. If our car is properly positioned, we'll be aiming just to the right of the entrance gate as we pass the apex. Some people put their inside wheels up on the yellow curb through West Bend. It's smooth enough that you can do it, but there's really little advantage unless you're avoiding traffic.

Wet: In the rain, a puddle can form just past the apex of West Bend, where there's a dip. You want to stay slightly outside it. The rough laddering in the corner will actually help you in the rain by giving better drainage than the polished portions of the line.

46.0
90 MPH

Dry: We let our car go out towards the edge, using all the width of the track. We want to be parallel to the edge of the track before we get to the Toyota Trucks Bridge. This is another place where you don't want to drop your left wheels off the edge and hook because you'll head right across the track and into the infield bank. You can see the dark tire marks where other drivers have done this. The simple remedy? Late apex West Bend.

Wet: You really have to watch out for wheelspin on the exit from West Bend, because you are on the normal dry line which is often soaked with oil and grease that comes to the surface in the rain. Particularly in a car with a lot of power, you have to be very smooth through here when it's raining.

47.0
92 MPH

Dry: This is a good spot to check the corner workers at the flag station under The Bridge on the left-hand side. If something has happened over the brow of the hill, you can't see it until you've hit it. The corner workers should warn you. Typically, cars will spin to the inside of The Downhill if they have trouble, so staying on your normal line to the left is the safest alternative. On the other hand, this is one of the best passing spots at Lime Rock. Many drivers are slower down the hill than they should be; you can pass them easily to the right going under The Bridge and, if you have to, accelerate down the inside, brake for the apex of the right-hander onto the Main Straight and still carry enough speed to stay ahead of the car you've just passed as you both accelerate down the straight.

48.0
95 MPH

Dry: In our Skip Barber School Formula Fords, we normally do a slow upshift from Third to Fourth just past The Bridge. This serves as our "lift." Depending on your car and its gearing, you may have to shift later or brake before the turn-in point at the bottom of the hill. IMSA GTP cars, for example, accelerate so fiercely that most drivers must upshift followed by heavy braking, all in the short space between The Bridge and the turn-in point. Other drivers will merely "breathe" the throttle in the midst of the hill as a sort of "confidence" lift. In many slower but good handling cars, it is possible to go flat-out from West Bend to Turn One without lifting at all. In our Formula Fords, we shift, then it's full throttle from The Bridge all the way to Turn One.

49.0
98 MPH

Dry: The track is not straight fromThe Bridge to the turn-in point of The Downhill turn. It goes in and out more than 2 feet. But you want to drive in a straight line to the turn-in point, hit the white blotch of paint and then turn. If you follow the edge of the track, you'll waste time and also take a chance of dropping a wheel off the edge of the pavement and hooking to the inside.

Wet: In the rain, you want to go down the hill on the right hand side, do "the slide for life" across the normal line and end up on the outside. The danger is from other traffic sliding down the hill on the normal line, at the edge of control. You will have to cut across their path to reach the outside of The Downhill, which is going to confuse them, if nothing else.

51.0
102 MPH

Dry: This has always been called The Downhill Turn. It's *not* a downhill turn. The turn is *after* you go down the hill, and the track actually goes slightly *up* as you come out of the turn, just beyond the apex. You can turn very abruptly at the turn-in point, because you're getting compression on the suspension as you hit the bottom of the hill and level out. There's a dip at the turn-in point that you want to avoid. You have to leave yourself a little margin to the outside.

Wet: When it's raining, the best traction on the whole race track is the outside of The Downhill. You take a diagonal line towards those signs on the outside, brake across the flat, trail on the brakes, brake hard when you land, trail off the brakes and hit the berm. Puddling is a big problem in the braking zone, though.

52.0
100 MPH

Dry: See how far around the corner the cone marking the apex is? If you early-apex this corner, you can end up in the stream. It has been done! It's very important to use all the track and take a very late apex. Within reason, the later the better. Taking a late apex lets you accelerate sooner onto the Main Straight. This is definitely a "Type One" corner. Since it leads onto the longest straight, it's also the most important corner at Lime Rock. If you do it right, you will carry your speed all the way down the hill, through the corner and down the straight till you brake for Turn One. If there's a secret to going fast at Lime Rock, it's this turn.

Wet: In the rain, you'll be on the far side of the normal line, then cut back across it as you start down the Main Straight.

53.0
100 MPH

Dry: We've hit the apex. We can also hold the car in along the yellow curb for a little to extend our apex. Whatever you do, don't lift abruptly. You'll be gone. If you get a little loose around the apex, don't over-react. Drive it out. If you get wide of the apex, let it stay wide and run it wide on the outside. Don't tuck tighter into the corner! You'll get yourself into Big Trouble by turning abruptly. You can feather the throttle a little bit, but if you lift, you've got a problem. Usually, you'll spin to the inside, into the guardrail. Another problem is that some guys will deliberately drop a wheel off on this corner, spreading dirt and gravel right on the line. Almost every weekend, somebody spins because the guy in front of him dropped a load of stones under his wheels.

56.1
101 MPH

Dry: This corner goes slightly off-camber to the left past the apex. If you drop two wheels off, you're in trouble because you'll hook back across and into the guardrail on the inside. If you jam on the brakes, you'll slide straight off the outside and into the tire wall. There is really no good way to leave the track in this corner. You want to use the full width of the track, but no more.

Wet: In the rain, you want to stay to the inside on the straight. There are three things to worry about in the rain. One is where to drive the car. Another is how to set up the chassis. The third is how to retain the ability to see. I've only been scared three times on a race track; twice my brakes failed, once in the rain at Watkins Glen my face shield fogged up, I couldn't get the snaps open with gloves on, and I couldn't see.

1:00.0
106 MPH

Dry: Now we're back on the Main Straight. In some cars, depending on gearing, you may have an upshift here into Top gear. This is a good place to check your mirrors for overtaking cars, to check your gauges, to check your pit crew for messages and the Chief Starter's flags. Usually, you'll try to pass on the right, which puts you on the inside going into Turn One. If you have the overtaking speed, this is obviously the easiest place to pass on the whole course. Your speed on the Main Straight, of course, is dependent on your speed through The Downhill. Inexperienced drivers often get stuck behind a slower car because the slow car holds them up in The Downhill and they don't have the momentum to overtake on the Straight. The solution is to outbrake the guy into Big Bend.

Butch Sherwood and the checker flag, enduring symbols of Lime Rock Park for thousands of loyal race fans.

(The following poem was originally published in an SCCA Northeast Division Regional newsletter in 1966 and reprinted in the Lime Rock Park program during 1980.)

...IT IS WINTER
Under a leaden sky
The wind rustles the last
Of the dead leaves in
The pits at Watkins Glen.
It is winter.
The cars are all gone now
To gather dust
In thousands of garages
From LA to Westport.
It is winter.
The snow goes unplowed
On the straights
Of Road America;
The empty Brut stands
Stare blindly about
At silence.
It is winter.
The girls are gone
Pounding typewriters in New York
Chicago and Baltimore
Wearing wollen skirts and pattern stockings
Instead of stretch pants and halters.
It is winter.
The brave young men are gone too
Talking sales and production
And costs and income
Instead of lines and drifting and oil pressure.
It is winter.
The unnatural silence
In the paddocks of Lime Rock and Thompson
Is deafening.
The Firestone and Goodyear trucks
Have fled.
It is winter.
Gone are the acres of spectators
And tents dewy-wet in the morning.
Stilled is the shouting and singing
And clank of beer cans in the night.
It is winter.
The last race is long since over
The last car trailered away
The last flagman gone home
The last gate padlocked.
It is winter.
The PA speakers speak no more
But the wind howls as it drifts snow
Around the empty rest rooms
And shuttered tech buildings.
It is winter.
The time for auto shows and rugged rallies
New cars and parties
And watching movies and slides
Getting high at the SCCA meetings
And planning and building
And waiting for next season.

—Pete Fullam

Index